# THE LASER THAT'S CHANGING THE WORLD

# THE LASER THAT'S CHANGING THE WORLD

## THE AMAZING STORIES BEHIND LIDAR, FROM 3D MAPPING TO SELF-DRIVING CARS

TODD NEFF

59 John Glenn Drive
Amherst, New York 14228

Published 2018 by Prometheus Books



Cover design by Jacqueline Nasso Cooke
Cover image courtesy of Harris Corporation
Cover design © Prometheus Books

Inquiries should be addressed to
Prometheus Books
59 John Glenn Drive
Amherst, New York 14228
VOICE: 716–691–0133 • FAX: 716–691–0137
WWW.PROMETHEUSBOOKS.COM

22 21 20 19 18 5 4 3 2 1

Library of Congress Cataloging-in-Publication Data

Names: Neff, Todd, 1968- author.
Title: The laser that's changing the world : the amazing stories behind lidar, from 3D mapping to self-driving cars / by Todd Neff.
Description: Amherst, New York : Prometheus Books, 2018. | Includes index.
Identifiers: LCCN 2018019450 (print) | LCCN 2018029341 (ebook) | ISBN 9781633884670 (ebook) | ISBN 9781633884663 (pbk.)
Subjects: LCSH: Optical radar—Industrial applications. | Optical radar—History. | Optical radar—Social aspects.
Classification: LCC TK6592.O6 (ebook) | LCC TK6592.O6 N44 2018 (print) | DDC 621.3848—dc23
LC record available at https://lccn.loc.gov/2018019450

Printed in the United States of America

# CONTENTS

# CHAPTER 1
# HUTCHIE

Flat-screen monitors display pointillist renderings of mountains, rivers, roads, bridges, buildings, and power lines with little regard for the usual relationships between actual pigmentation and color. In place of those are schemes generally favoring deep blues at the base that then loosen into greens, yellows, oranges, and reds with increasing elevation. There are enough screens with enough color that one has the sense of being courted by many amorous peacocks. The screens belong to exhibit hall booths, which, for the duration of the 2017 International Lidar Mapping Forum at the Hyatt Regency Denver, belong to dozens of data providers, point cloud software developers, and makers of very pricey laser and other hardware. Many have sci-fi names: Quantum Spatial, Network Geomatics, Terrasolid. Many have sci-fi products too, though they happen to be real.

Drones resembling paper airplanes, stealth fighters, helicopters, and propellered crabs have alighted on many tables and stands. I pause to consider the Pulse Aerospace Vapor 15, a black chopper with its rotors folded back like the wings of some immense cricket. If it were a carnivore, I would be lunch.

I just had lunch, standing over a ham and cheese sandwich from the buffet with a semiretired guy named Brad Weigle. He grows bananas and mangoes in his Florida backyard but won't be there much over the next couple of years. He's moving to Hawaii to map the National Tropical Botanical Gardens in high resolution using lidar that flies on drones like the Pulse Aerospace Vapor 15. As we talked, Chuyen Nguyen, a graduate student from Texas A&M–Corpus Christi whom Weigle had met earlier, stopped to say hi and somehow mistook me for someone capable of grokking something called multiscale voxel segmentation for terrestrial lidar data as pertains to swamp mapping, the subject of her PhD work. I smiled and nodded pleasantly and said things like "Wow" and "Amazing" at what seemed be appropriate intervals. When she moved on, Weigle assured me: "That stuff is cutting edge. It's going to be a really big deal."

Over at the Velodyne Lidar booth, a sales engineer named Jeff Wuendry tells me that self-driving vehicles that constantly lidar-scan their environments (so they, for example, don't smash into things) could, if that data were crowdsourced and stored in the cloud, obviate the need for lidar mapping drones, at least in a lot of urban environments. Wuendry also tells me that he used to work for a German company called SICK, now a big name in lidar. It started out with light curtains so metal-stamping machines wouldn't inadvertently crush the hands of their minders, he says.

I walk the few steps over to the Harris Corporation booth, where Blake Burns, a senior sales engineer, explains that with one of Harris's multimillion-dollar Geiger-mode flash lidar rigs, from an aircraft flying 330 miles an hour at an altitude of twenty-seven thousand feet, can render a three-mile swath of whatever lies below in photographic detail (rainbow coloration notwithstanding)—not to mention, he adds, three-dimensional exactitude to about four inches in any direction. The US military has been using similar Harris systems for two decades, he says, though this was only declassified a couple of years back. Just three of the civilian units exist, and they have better things to do than hang out in an exhibition hall.

Right next door, so to speak, Katie Fitzsimmons of Leica Geosystems tells me that Leica's single-photon lidar—another multimillion-dollar black box, an example of which is on hand and, were it turned on, would be collecting breathtakingly detailed imagery and elevation data of the Hyatt carpet—operates like the equivalent of one hundred typical lidars at once, firing off six million laser pulses each second. The company is in the process of elevation mapping the continental United States and Western Europe to an accuracy of a foot or less with these machines, Fitzsimmons adds. She speaks with enthusiasm, but also more matter-of-factly than the situation seems to warrant, like someone reciting for the umpteenth time the technical specs of a space-time portal.

Her colleague Josh Rayburn describes the company's Pegasus:Backpack, a twenty-eight-pound carbon-fiber-reinforced wonder with five cameras and two spinning Velodyne lidars so someone can walk around and capture detailed 3D renderings of indoor or outdoor spaces. It costs about as much as a Ferrari. It's right about now that my brain starts to hurt.

It's not only being overwhelmed by all the shiny objects, preening flat screens, and jargon; it's also the weight of knowing that the lidar mapping that this entire conference is dedicated to represents just one of many realms in which people are using lidar. That knowledge both reinforces and seriously complicates my desire to make the technology and its evolution clear to people who, like me, might otherwise suspect a voxel to be a Toyota subcompact.

Lidar is a technology capable of measuring continental drift, determining the composition of Earth's atmosphere, discovering lost cities, tracking the biomass in forests, assessing flood risk and hurricane damage, measuring the melting of glaciers, detecting submerged explosives and the level of the seas in which they lurk, listening to stinkbug conversations, guiding missiles and self-driving cars, and a whole lot more. I have become fascinated with this most striking macroscopic application of quantum physics, which, were it not for its milquetoast name, would surely have soaked up a lot more popular love by now, considering its growing importance to science, industry, and government. I have decided it's time to tell lidar's story, starting at the beginning and hanging tough until a sort of techno-Darwinian explosion sent the technology radiating into more niches than any reasonable tome might possibly hold. Then I will have to pick and choose.

The beginning is straightforward, though. The extraordinary technologies arrayed at this technical conference in the Mile High City share an improbable common ancestor: lidar's first flicker occurred between the ears of a self-educated, mentally troubled Irish savant who lived with his parents on the outskirts of Dublin and corresponded with Albert Einstein. He dreamed up ways to see the impossibly tiny and the impossibly distant, to be realized long after he was gone.

Edward Hutchinson Synge, known as Hutchie, was born in 1890 to a well-to-do family in Dublin whose best-known member was Hutchie's uncle Johnnie, with whom he was particularly close as a boy. The playwright John Millington Synge had been a leading light of the Irish Literary Revival, known for his no-punches-pulled depictions of the Irish peasantry and his willingness to lay bare the ironies of the Irish psyche. There were full-on riots when his play *The Playboy of the Western World* opened in 1907 at the Abbey Theater in Dublin, which J. M. Synge had cofounded with his old friend William Butler Yeats, among others.

Hutchie was educated at home by governesses and tutors until he enrolled in mathematics and Gaelic—a passion of his famous uncle—at Trinity College Dublin in 1908. He was a brilliant student. When Hutchie's uncle Johnnie died of cancer at age thirty-seven the following year and left Hutchie nearly half his estate, the younger Synge abandoned his degree a year shy of graduating to pursue a life of independent scholarship. His uncle, who had once done much the same thing, was the inspiration.[1]

It could have been that Hutchie felt hemmed in by prescribed study, or that he understood he would never quite fit in academia. He was dark-haired, handsome, tall, and, like his rugby-playing younger brothers, sturdily built. But he had no interest in sports or a social life, preferring books to beings. He was shy and quiet and could come off as arrogant, and he could do absolutely nothing about it.

Today, Hutchie in all likelihood would have been diagnosed with Asperger's syndrome, which can yield stratospheric IQs and obsessions with complex topics, but also emotional distance and poor social skills. In this respect, as would also prove to be the case with his scientific insights, Hutchie was born decades too early. There was no cognitive behavioral or talk therapy to teach the young Hutchie how to interact with others and interpret intellectually (rather than instinctively) things like facial expressions and tones of voice. Back then, mental and behavioral disorders were handled quietly within the walls of the family home, and so it had been with Hutchie Synge.

Hutchie was in good company, at least. His uncle's friend Yeats probably also had Asperger's, and J. M. Synge himself may have been somewhere on the spectrum.[2] Perhaps the playwright recognized it in the nephew, making his bequest as a buffer against the difficulties in life Hutchie would doubtless encounter. Or maybe he recognized a kindred spirit in an altogether different sense. As Hutchie later recalled, his uncle Johnnie held "very odd and altogether ridiculous views upon me, believing me to be a changeling whom the happenings of his own life had somehow or other brought into existence . . . a sort of double being."[3]

Either way, Hutchie left Trinity. As his brother John Lighton Synge, an eminent mathematician and physician, put it many years later, their famous uncle had inadvertently "deprived the academic world of perhaps the best Irish mind of the century, certainly the most wide-ranging."[4] But that mind, too, would grow increasingly troubled. The term had yet to be coined, and it's hard to diagnose across decades, but all signs point to bipolar disorder that grew more corrosive with time.

**Edward Hutchinson "Hutchie" Synge. (Photo courtesy of Living Edition Publishers, Austria.)**

Over the next couple of years, Hutchie traveled, as his uncle Johnnie had done. Perhaps he went to Germany, probably he went to France, certainly he visited the Aran Islands off Galway—but no evidence survives. By 1916, as World War I raged on the continent and the Irish War of Independence started at home, Hutchie had returned to Dublin. He read extensively: as his physicist brother later described it, Hutchie built up a personal library that spanned the classics, art, literature, philosophy, and science, reading avidly and retaining most of it. "I never met anyone who gave me such an impression of omniscience," J. L. Synge wrote.[5]

By the early 1920s, Hutchie had anointed himself as de facto literary executor for his uncle's works, which he had intended to augment with an authoritative biography of his own creation. His efforts included enlisting Yeats's help in tracking down letters J. M. Synge had written a young woman who had declined his hand in marriage. The woman, by then married and no longer young, lived in Cape Town.[6] Despite precarious mental and physical health that put Hutchie, at various times, bedridden in hotels and nursing homes, Hutchie traveled to South Africa, presumably to meet the woman himself.[7]

By then, he had been drawn to physics, a subject he had never formally studied. In particular, this college dropout became obsessed with the idea of publishing the collected papers of the great nineteenth-century Irish physicist, astronomer, and mathematician Sir William Rowan Hamilton. Hutchie was not alone in considering him "one of the greatest—if he was not indeed the greatest—mathematician in Europe during the past century." He was aghast that the great man's publications had never been consolidated, that the ores of two-hundred-plus notebooks, all in storage in the Trinity College library, were still waiting to be mined more than sixty years after Hamilton's death. "Hamilton has been treated by his country, and by the learned bodies in Ireland with which he was associated, with a neglect which no contemporary mathematician of any other European nation, who at all approaches his rank, has suffered," Hutchie wrote.[8]

Lacking the academic credentials demanded by his projects, he sought support from high places, going as far as writing to Albert Einstein himself. In May 1922, the great physicist responded, confirming Hamilton's huge influence on physics as well as on Einstein's own special relativity work. Einstein concluded that a one-stop shop for Hamilton's wide-ranging publications would be "of great importance."[9]

The extent and thoughtfulness of Einstein's reply to an unknown correspondent is remarkable—all the more so given that by the time he wrote to Einstein, the thirty-two-year-old Hutchie had published misguided assaults on special relativity in letters to the journal *Nature* and in a report for *Philosophical Magazine*. "The doctrine of relativity, as such, therefore breaks down," Hutchie had concluded.[10] However, it's unlikely that Einstein ever saw it: when you call into question the universal applicability of Newton's two-hundred-year-old laws of motion, skeptics emerge at the speed of light.

Hutchie's Sir Rowan Hamilton project stalled, though, when the two Trinity College mathematicians he had been cultivating to take it on died in quick succession. His work on the biography of his uncle also hit a wall. His health wasn't great, either: in early 1925, Hutchie mentioned to a correspondent that he had been bedridden for three months—likely due to a depressive episode.[11] His brother J. L. Synge, back from the University of Toronto to take a faculty position at Trinity College Dublin around that time, stepped in.

"The idea of his writing up his work came from my father," recalls Cathleen Morawetz, J. L. Synge's daughter. "Hutchie was in a rather despondent state, and my father suggested to him that he should write up his discoveries." What would become Hutchie's unlikely contributions to science went from Hutchie's room in his parents' house on Sydenham Road in Dundrum first to his brother, then to John Joly, whom Morawetz described as "one of the most important physicists in Ireland at the time."[12] Joly was an old family friend as well as an editor of the respected London-based *Philosophical Magazine*. Hutchie's first idea involved seeing invisibly tiny things.

Six years after their first correspondence, Hutchie wrote Einstein again, sending a seven-page handwritten letter to Berlin on April 22, 1928. Hutchie led with the good news: his brother J. L. Synge and a colleague, thanks to Hutchie's relentless prodding, were at work on Hamilton's papers. He then transitioned to a seven-page technical design for a new sort of optical instrument. "I hardly know whether what I am enclosing may interest you," Hutchie wrote. "It is the outline of a method which has occurred to me, and which confirms the principles of the ultramicroscope."[13]

He envisioned a microscope that moved a tiny aperture back and forth while extremely close to the object being studied. It would bring into focus things much smaller than the wavelengths of light shone on them. This was a radical idea: more than a half century earlier, Ernst Abbe had established his diffraction limit. It held, in essence, that a microscope will blur the image of anything smaller than the wavelength of the light being used to look at it. That meant beyond a lower limit of about 250 nanometers—billionths of a meter—all would remain forever invisible, leaving things like viruses and structures inside of cells forever cloaked in mystery. Yet here was Hutchie, begging to disagree from his quiet room in Dundrum. After consulting with others he deemed "qualified to express an opinion either on the theoretical or practical side," he concluded that his method was "theoretically sound, and that the practical difficulties—while formidable—are not beyond present day technical achievement."[14]

This was all a long way from the theoretical trails Einstein had blazed to global fame. But then, Einstein had written his 1905 papers on the nature of light—including the iconic $e = mc^2$ equation—while spending his days as a patent clerk considering the viability of various notional contraptions. What's more, he had recently coinvented a refrigerator that ran on a heat source.[15] Einstein's epochal gifts for grasping the architectures of creation extended to the constructs of humankind, and Hutchie would benefit.

The great physicist responded to Hutchie's letter eleven days later, in typewritten German. He expressed his delight in the Hamilton publication efforts and then turned a critical eye on Hutchie's design. In general, he agreed with it. Einstein had questions, but the great scientist admitted to difficulty in decoding Hutchie's handwriting too: "Maybe I did not understand you correctly, but there was so much in your letter that I could not read, I had to guess—that it is quite possible—that your letter nevertheless contains the solution of the task. I would be delighted to hear your opinion on this," Einstein wrote.[16] That Einstein could recognize the kernel of originality in a largely indecipherable letter from a complete stranger says as much about the man's patience as it does about his brilliance.

Hutchie answered quickly with clarification, and Einstein responded again on May 14. Following a bit of technical advice and correction, Einstein concluded, "It might be most proper to publish the idea as such in a scientific journal and at the same time to point to the practical difficulties in its realization," including, as an example, the need for a precise aperture just one millionth of a centimeter wide, for which there was no conceivable approach at the time.[17]

Less than two weeks later, Hutchie sent a paper, "A Suggested Method for Extending Microscopic Resolution into the Ultra-Microscopic Region," off to his brother and, ultimately, *Philosophical Magazine*. What impact Einstein's blessing had on John Joly and colleagues' decision to publish the paper, we can't know. Nor is it clear to what degree the famous scientist's encouragement emboldened Hutchie, who by then was thirty-eight years old and had little to show for all his scientific self-study and rumination. But the ultramicroscope paper sparked four years of creative output that would make up for his late start.

His next effort would prove no less visionary. This time, Hutchie looked to the heavens. He focused not on the obvious—the clouds down low, the sun and the stars on high—but the invisible stuff between. The idea was to use an array of searchlights to study the upper atmosphere, which he felt would be "of considerable importance to meteorologists" due to possible interactions of the stratosphere with the weather we all feel.[18] His approach would work far above the twenty-kilometer (twelve-mile) ceiling of balloons, to as high as fifty kilometers (thirty-one miles), he said.[19]

The observer would aim the searchlights skyward, at night, from a mountaintop. Perhaps eighteen miles away, there would be a meter-wide reflector that bounced light into a photodetector (like the eye's retina or a digital camera sensor, a photodetector converts light into electrical signals). With the searchlight beams shining skyward at a fixed sixty-degree angle, one could aim a telescope at various heights of this artificial, inverted Jacob's ladder, figuring out the altitude based on the angles. The searchlight beams themselves would bounce too little light off the sparse atmosphere at high altitude for the human eye to perceive. But the reflector and photodetector, Hutchie calculated, could manage it as long as there were enough searchlights. So one could, he figured, determine how many molecules were zipping about at a given altitude—the atmospheric density—without actually sending a scientific instrument to great heights. In modern scientific argot, Hutchie was proposing the first use of active electro-optics for atmospheric remote sensing. But more on this later.

This would be big science, a 1930s version of a particle collider or a space telescope. To generate the colossal volume of lumens needed and capture changes in the upper atmosphere over time, "a permanent assemblage of several hundred searchlights would be required," Hutchie wrote—searchlights that, including the massive diesel generators, cost more than $800,000 each in today's money. But that part was less outlandish than at first blush: there were lots of searchlights left over from World War I, during which, among other things, they had helped spot marauding German zeppelins over London. "This offers no great difficulty," Hutchie said, "since all the belligerent nations in the late war have many thousand searchlights on their hands."[20]

As with his ultramicroscope idea, he paced through the technical hurdles in his mind's laboratory: the challenges of calibrating the distant detector; of ensuring consistent light quality despite vagaries of the carbon arcs blazing away in the searchlights (having a lot of them would be a good thing, Hutchie argued, as the variability would average itself out); of compensating for nettlesome starlight; of needing an extremely smooth and dazzlingly shiny reflector. As with his ultramicroscope idea, Hutchie was optimistic, to put it mildly: "Nearly all the apparatus required exists in sufficient quantity; there are no formidable technical difficulties to be surmounted; and the theory appears quite unassailable."[21] In May 1930, *Philosophical Magazine* published the idea under the title "A Method of Investigating the Higher Atmosphere." With this, Hutchie introduced an entirely new approach to measuring things in the sky. Until that point, conventional wisdom held that to measure something in the sky, one had to loft some sort of sensor to altitude. Hutchie's hardware could stay on the ground.

Unlike the ultramicroscope idea, which attracted little interest, others took note this time. By 1935, researchers from the Carnegie Institution in Washington and the US Department of Agriculture had teamed up to run some numbers and do lab experiments. They concluded that, with a few tweaks—using mechanical shutters to sort of Morse-code the beam to make it more distinguishable to a more sensitive receiver—the scheme might work for altitudes of seventy kilometers (forty-three miles), maybe even higher.[22]

Two years later, on a moonless night in April 1937, Edward O. Hulburt, a US Naval Research Laboratory scientist, put Hutchie's ideas to the test using an actual searchlight. Hulburt persuaded the US Army Corps of Engineers to fire up a six-cylinder generator and point their sixty-inch, eight-hundred-million-candlepower carbon arc illuminator (roughly the brightness of the headlights on six million cars) into the skies over Virginia. Eleven miles to the northwest, Hulburt set up a camera with sensitive film and took a series of long exposures. Looking closely at the photos from this night and a follow-up a month later, he found that amid streaks of star motion, the film picked up traces of the searchlight up to twenty-eight kilometers (seventeen miles)—more than a mile higher than the Explorer II stratosphere balloon the National Geographic Society had launched in 1935, he noted—and that a more intense beam and a mountaintop location could extend the technique's reach higher yet. Just as important, Hulburt calculated atmospheric scattering at increasing altitudes based on what he saw, and it more or less jibed with what theory would have expected.[23]

The Carnegie Institution took the baton again in fall 1939. Ellis A.

Johnson, who had worked on Carnegie's 1935 lab experiment, led the group. They designed their own carbon arc searchlight with a two-foot mirror loaned from the Sperry Gyroscope Company, housing it in a welded-pipe frame supported by a platform of wooden four-by-fours. They tipped the rig sixty degrees to the north and anchored it with a length of rope. Atop Johnson's searchlight were metal shutters like the ones they had tested in the lab four years earlier. They connected to an electric motor that opened and closed the shutters to "fingerprint" the light for the receiver.[24] Over a few clear, moonless nights that August and September, Johnson complemented his business suit with a welding mask to shield his eyes from the glare of his creation.

Johnson also spent time three and a half miles north, in Kensington, Maryland, where he aimed the receiver at different points along a beam mostly invisible to the naked eye. The receiver had a thirty-inch mirror, a photomultiplier to boost the faint signals, and two Bell Laboratories–supplied photoelectric cells. The combination captured searchlight signals as high as forty kilometers (twenty-five miles). Above the city's low-lying haze, scattering theory and the experiment's findings fell into lockstep. Johnson also found that "many apparently clear nights actually proved to be considerably disturbed by the presence of invisible clouds and absorbing material," particularly from ten to twelve kilometers (six to seven miles) in altitude. "Other nights showed the presence of a varying absorption or scattering of a fairly rapid period," he found.[25] The significance is easy to miss: from the ground, Johnson was capturing the fickleness of the stratosphere in real time.

With some upgrades (bigger searchlight and receiver, greater distance between searchlight and receiver), Johnson figured he could observe the atmosphere as high as seventy to eighty kilometers (forty-three to fifty miles) using the technique. In addition, much more so than Hutchie or Hulburt seemed to have done, Johnson grasped the breadth of atmospheric riddles Hutchie's inspiration might solve: the height and amount of water vapor, turbulence, winds, dust, and ozone. With ozone in particular, Johnson realized that the approach, with a few adjustments, could establish ozone distribution based on how that particular molecule scatters a certain band of light.[26] In other words, you could use light to learn not only how many molecules were floating about at a particular altitude but also what kind of molecules they were.

With hardware in hand and a clear idea of how further inquiry into Hutchie's idea might unfold, Johnson seemed poised for a breakthrough. But even as he was pointing his searchlight into the sky, Germany was invading Poland. He ended up working on underwater mines and countermeasures for the US Navy during World War II. Then he got interested in operations research, became a pioneer in the field, and, it's safe to say, never blasted the sky with light again.[27]

Once Hutchie had published his searchlight idea, he, too, refocused. In his case, he looked well past the atmosphere to invisibly distant things. In August 1930, just three months after his searchlight idea saw print, he published two more papers in *Philosophical Magazine*. Both dealt with new ideas for astronomical instruments. One had to do with a redesign of an optical device called a Michelson interferometer; the other was "A Design for a Very Large Telescope."

Hutchie envisioned a telescope in which "a number of moderate-sized parabolic reflectors perform the functions of a single, very large reflector. Theoretically, at least, this form of telescope could be made to give virtually the same degree of resolution as a telescope with a single large reflector."[28] He proceeded to walk through the benefits of constructing a telescope that is in fact many telescopes: better resilience to temperature changes, less weight, cheaper. Hutchie also noted that such a telescope could well exceed the practical limit of a ground-based telescope's resolution—which in the days before adaptive optics was limited by wind turbulence—by setting it up in, say, Antarctica. There, he noted, the British Antarctic Expedition of 1910–1913 observed that smoke from Mount Erebus "rose so vertically that no direction could be assigned to it."

"It is not, therefore, of merely theoretical interest to show that the type of telescope suggested is available for great resolving power," Hutchie said, assuming, of course, "perfect enough" workmanship and adjustments.[29]

As had been the case with his ultramicroscope idea, Hutchie's imagination far outstripped the practical capabilities of the day, and the idea found no traction. Hutchie would send four more papers to *Philosophical Magazine*, touching upon the galactic (determination of stellar parallax using an interferometer like the kind he had recently proposed; a discourse on why stars twinkle leading into a slight modification of his very large telescope design) and the microscopic (two refinements to his ultramicroscope notion). The last of these appeared in 1932. It would be the last thing he ever published.

There was a falling-out between the Synges and John Joly, and then Joly died in 1933. Probably Hutchie's declining mental health was also decisive. J. L. Synge, back at the University of Toronto by 1930 and concerned with his brother's state, urged his daughters to write Hutchie letters as some form of human connection. The first few sentences of Hutchie's replies came across entirely as one might expect in a correspondence between an uncle and his young nieces. Then his writing would take flight like leaves swept up in a gust, with rantings on random topics occupying his formidable, omnidirectional mind—the status of German President Hindenburg, say—marked by obsessive underlining, as Morawetz recalls. Her father, J. L. Synge, later mentioned notebooks with similar patterns of lucidity and mania.

The death of Hutchie's mother's in 1935 eroded his already-precarious mental health. In 1936, there was a blowup of some sort, Morawetz says, maybe involving a knife and threats to his older sister Ada, who was by then running the Dundrum household. Hutchie's physician brother Millington diagnosed "persecution complex" and committed him to Bloomfield Hospital in the Donnybrook area of Dublin.[30]

His many notebooks stayed in Dundrum. Hutchie, writing later from Bloomfield, feared they would be destroyed, and his fear was warranted.[31] Neither those he left behind nor the ones he went on to write in Bloomfield survive. What insights into the ways of light were lost?

Cursory monthly reports from his minders offer only glimmers of insight into his twenty-one years there: that he had good days, in which he read avidly and behaved as the courteous and educated gentleman he was; that he had less good ones, in which he looked awful, stayed in bed, and suffered paranoid delusions; that his appeals for release were always denied. When he was well enough to be allowed out during the day, he took solitary walks in Dun Laoghaire on Dublin Bay and went to the movies—alone, as he always had been. So it was until May 28, 1957, when a stroke felled him.[32]

Hutchie never did finish the biography of his uncle. But he did shepherd the 1927 publication of a limited-edition version of *The Playboy of the Western World*, one distinguished by ten illustrations by Seán Keating, by then well-known for his depictions of the Irish War of Independence. According to J. M. Synge biographer William McCormack, it was through this book that, nearly two decades after the playwright's death, "Synge's reputation would be revived."[33]

Hutchie's enlistment of his brother to work on the Hamilton papers turned out to pay unexpected dividends too. John Lighton Synge finished the first volume of Hamilton's collected works in 1931. "I became fascinated by this intimate contact with a great mind, which ranged with ease from abstract idea to very detailed arithmetical calculations, in which I was never able to find an error," the physicist later wrote. He would go on to write two of his own books as a direct result of the project, and scores of his papers involved Hamilton's methods, some focusing on them exclusively.[34]

Hutchie's ultramicroscope idea came to be in 1984, more than a half century after he had conceived it. That's when IBM scientists invented the scanning near-field optical microscope. It could resolve details just twenty-five nanometers across—three thousand times smaller than the diameter of a human hair—despite using light with a wavelength twenty times longer.[35] Its inventors had never heard of Hutchie or his work, but he had foreseen their solution, and he is now recognized as the father of the field. The technology continues to develop, with the devices shedding their precise light on nanowires, cellular structures and proteins, the surfaces of solar cells and optical switches, and other infinitesimal building blocks of natural and human invention.[36]

Hutchie's telescope design would be realized decades later as well, on a mountaintop in southern Arizona. Scientists combined into a single unit six telescopes, each with a six-foot mirror once destined for a spy satellite the US Air Force never launched. The Multiple Mirror Telescope (MMT) opened its eyes to the skies above Mount Hopkins in 1979. "The paper by Synge (1930) is especially fascinating because it appears to be the earliest description of an MMT-like device," astronomers wrote. "His description of ideas for the optics and the support arrangement for the optics in fact resembles the MMT so closely that the MMT, although conceived independently, could be viewed as the realization of the Synge concept."[37] Hutchie's instincts about Antarctica as an astronomical perch proved right too. In 1994–1995, astronomers set up the first remote observatory on the southern continent. "The South Pole environment is unique among observatory sites for unusually low wind speeds, low absolute humidity, and the consistent clarity" of the sky, they wrote.[38] Others, most notably the South Pole Telescope, have since followed. As has lidar, as we'll see.

Hutchie's searchlight idea would prove no less prescient as it evolved and radiated into countless niches across diverse realms of science, commerce, and defense, a process that continues nearly ninety years later. But a particular enabling technology had to come first. That technology would be born a short while after Hutchie was buried at Dublin's Mount Jerome Cemetery, in a plot not far from the final resting place of his uncle Johnnie.

# CHAPTER 2
# ENTER THE LASER

Today's lidar depends on a lot of things: microprocessors, light detectors, digital storage, software, and sophisticated mechanical and digital means of sweeping beams of light across fields of view. It owes its existence to a happy convergence of many inventions, each with its own story that reaches back further than you might guess. (The first "digital" storage? Punch-card-like paper tape to control textile looms, circa 1725).[1] The laser at the heart of lidar has a history with roots deeper yet, dating back to ancient Greece and beyond. That's because to make a laser, scientists had to first understand what light really is. Then they had to master it.

What is this strange stuff that illuminates Earth and skies, in whose absence the most vivid colors go black? The question vexed philosophers, then scientists, for centuries. The ancient Greeks bandied about a few ideas. Some, including Plato and Euclid, embraced various takes on what we now call extramission, believing that light was in part or entirely a product of some sort of "visual fire" streaming out of our eyes. Aristotle, who offered up the first detailed discussion of vision, argued for the light coming in, or intromission. So far, so good. But then he added that light depends on a transparent medium whose role as a carrier is a lot like that of air in the transmission of sound. Just as there could be no sound without air, there could be no light without this medium—call it "luminiferous aether," though that term wouldn't emerge for another 2,200 years. It gives you an idea of how long a shaky idea can stick around. The color of an object, Aristotle said, manifests itself through the intermingling of light and this static aether between the object and the eye.[2]

There was skepticism even in Aristotle's time. While some of his predecessors and contemporaries agreed that light wasn't a product of human eyes, they didn't fully buy into the aether intermingling. Rather, they believed that light moved in straight lines at extreme or infinite speeds. It took a while for the tide to fully turn—until about the year 1000, when the Arab scientist Alhazen finally established through experimentation that light indeed moves really fast, and in straight lines. He also blew holes in the idea of extramission with such questions as a) If the eyes emit light, why does it hurt to look at the sun? and b) How could human eyes possibly emit enough light to illuminate all the heavens?[3] Extramission's 1,500-year run had ended, besides among those who believe they can "feel" when someone's staring at them.

Centuries again passed until scientists in Europe, with help from Alhazen's writings, seriously considered the nature of light. In the 1600s, some who became famous for other scientific contributions chipped in. Galileo used a couple of lamps to try to measure the speed of light, concluding that the technology wasn't quite up to the task. Johannes Kepler, known for his laws of planetary motion, introduced the idea of vision being a product of light striking the eye's retina. Using prisms, Isaac Newton discovered a rainbow of color hiding in the sun's white light; later, he became a champion of the idea that light is made up of particles. Christian Huygens, who had discovered Jupiter's moon Titan (and, as a side gig, invented the pendulum clock), was among those who subscribed to the competing wave theory of light. The two sides would duke it out until the early twentieth century, when Albert Einstein said they were both right.

Things started happening faster by the 1800s. John Dalton in 1808 took Robert Boyle's notion of chemical elements a step further and suggested that there were lots of kinds of atoms, each a different substance. In 1814, Joseph Fraunhofer, a Bavarian optics wizard, noted hundreds of dark lines in the solar spectrum. Forty-five years later, Robert Bunsen and Gustav Kirchoff determined that different atoms produced different spectral lines when heated (by—what else?—a Bunsen burner), discovering the elements cesium and rubidium along the way.

By 1865, James Clerk Maxwell had unified electricity, magnetism, and light, which made clear in math that visible light was just a slice of a much broader electromagnetic spectrum, most of which we can't see. (This wasn't necessarily a complete shock: William Herschel, discoverer of Uranus, also discovered infrared light in 1800. While measuring the temperatures of different parts of a crystal's rainbow, he noted that the heat increased as the measurement moved from blue through red and was greatest in the "dark" beyond the red.) In 1888, Heinrich Hertz proved the existence of the electromagnetic waves Maxwell's equations had insisted on. The oscillations of power plants and processor cores are named in his honor.

Meanwhile, the notion of Aristotle's mysterious medium—that luminiferous aether—hadn't gone away. If light was in fact a wave, it needed something to grab onto, right? In 1887, Albert Michelson, who had already pegged the speed of light at 186,300 miles per second (within 0.01 percent of modern measurements), teamed up with Edward Morley to characterize this aether and instead found absolutely no hint of it. The interferometer Michelson invented for the task was the same one that Hutchie later suggested tweaking in one of his *Philosophical Magazine* papers.

By 1900, scientists had figured out that the chemical elements Boyle proposed seemed to have patterns to them, which Johannes Robert Rydberg called "periodicity" (hence the *periodic* table). Within a couple of years after J. J. Thomson discovered the electron in 1897, Max Planck had come up with quantum theory, which says an atom may only emit or absorb energy in discrete packets he called quanta. In 1905, the twenty-six-year-old Einstein resurrected the particle side of the particle/wave debate. He deduced that light, despite clearly coming in waves from the point of view of reflection, refraction, and so on, "consists of a finite number of energy quanta, localized at points in space, which move without subdividing and which are absorbed and emitted only as units."[4] Without these quanta, there could be no lasers.

Niels Bohr in 1913 reconciled Planck's and Einstein's ideas, showing that orbiting electrons go into an excited state when they absorb a discrete packet of energy—and that they release a discrete packet of energy when their excitement wanes again. These packets of energy, quanta, come in the form of light (and its fellow electromagnetic radiation).

With that, humanity had, after eons of wondering what makes the sun shine and fires glow, finally arrived at the answer: electrons getting a bit less excited and casting off little packets of light as they plunge to a lower energy state. When it happens in nature, it's called "spontaneous emission." Spontaneous emission encompasses basically all the light we see, from the glowing of touch screens to the flickering bottoms of fireflies. But in 1917, Einstein proposed another possibility, one called stimulated emission. As with a lot of what happens in the quantum world, it seems nutty. It goes like this.

If an unexcited electron gets hit by a light packet—a photon—of just the right wavelength, the electron gets excited, jumping to a higher energy state. When it relaxes again and hops back down, it releases a photon of that same wavelength. That's your standard spontaneous emission. But what if an electron is *already* excited and gets bonked by a photon of that same just-right wavelength? You'd think it might get more excited. But no. Instead, the electron goes slack and sends off a photon of that same wavelength. And here's the kicker with stimulated emission: the first photon, rather than being absorbed by the electron, *just keeps going as if nothing happened, and in the same direction as the newly created one*. Where there was one photon, we now have two. If you pile enough of the atoms hosting these electrons together, get them excited, and then bombard them with just the right kind of light, you can trigger a photon cascade that happens to move along in the same direction, with the same energy, and with the crests and troughs of its waves in perfect synchrony. The combination is known as coherence, and it's what makes a laser a laser.[5]

Done. Except for the detail of actually making it happen, which would take another forty-three years and a world war.

World War II pushed ahead many technologies, but none more than radar. Heinrich Hertz himself had noticed that some surfaces absorbed his radio waves and others reflected them. Proto-radar came along as early as 1904, when the German scientist Christian Hülsmeyer patented what he called a *Telemobiloskop* capable of bouncing radio waves off ships as far as three kilometers (two miles) away.[6] But the British pioneered modern radar. Robert Watson-Watt, a descendant of steam engine pioneer James Watt, had worked on systems to detect thunderstorms based on the radio waves lightning produces. In 1934, the Air Ministry asked him to check into the possibility of Hitler's Germany actually making a rumored radio "death ray." Watt found that to be fanciful but wrote a memo about his recent work in "the difficult, but less unpromising, problem of radio-detection as opposed to radio-destruction."[7]

He had a patent by the next year, and his system became the basis for the Chain Home radar stations credited for tipping the scales for the British in the Battle of Britain of 1940. But the radio waves' roughly ten-meter (33-foot) wavelengths demanded huge antennas and made it hard to pinpoint enemy aircraft (the longer the wavelength, the bigger the antenna and the fuzzier the picture). Another British invention that same year shortened the radar's wavelength by a factor of one hundred—into the microwave range—and made radar small enough that it might fit in an aircraft. This device was called a cavity magnetron, and you're well familiar with the technology, if not the name: it's the black magic in your microwave oven.[8]

The Brits needed more of these cavity magnetrons than they could possibly build. In September 1940, they sent a delegation to Washington with a ten-centimeter magnetron for some top-secret show-and-tell. The Americans, who had been working on microwave radar with less success, took little convincing to establish the Massachusetts Institute of Technology Radiation Laboratory, which was up and running before the end of that year. By the war's end in 1945, an MIT Rad Lab staff that peaked at about 3,500 people, with many industry collaborators, had developed more than a hundred radar systems: airborne radars, shipboard radars, coastal defense radars, fire-control radars, antisubmarine radars, and many more. The atomic bomb ended the war; radar may well have won it.[9]

The work left an enormous talent pool, one that kept pushing microwave technologies into shorter and shorter wavelengths promising smaller systems, sharper identification of objects, and faster communications, among other benefits. The US military, interested in all these things and more, paid for a lot of it. Over the next few years, scientists managed to create shorter and shorter microwaves—within about a decade, magnetrons had squeezed signals down to 2.6 millimeters (a tenth of an inch), crest to crest. But they could go no further.[10] They also proved out Einstein's theory of stimulated emission with the maser, or "microwave amplification by stimulated emission of radiation." The first was built by a Columbia University team led by Charles H. Townes, a radar expert, in 1953. Townes and a slew of other scientists then set their sights on doing the same thing with light.

Why? At frequencies roughly a hundred thousand times higher than microwaves, lasers ("light amplification by stimulated emission of radiation") could enable high-speed optical communications. Their tight beams could provide much more detail about incoming aircraft or other distant objects. They could be used for superaccurate targeting and perhaps even as enablers of nuclear fusion or directed-energy weapons, not to mention the scientific applications in areas such as spectroscopy—understanding what something is made of based on the light its electrons emit or absorb.[11] Tattoo removal, eye surgery, supermarket scanners, digital highways, the pointers cats love to hate, and so many other uses waited beyond the horizon of sane prognostication.

The bigger question was *how*. Light is, photon for photon, thousands of times more energetic than microwaves, and light's far shorter wavelengths rendered the approach Townes and others had taken with the maser untenable. Townes and Bell Laboratories scientist Arthur Schawlow came up with an alternative at about the same time Columbia University graduate student Gordon Gould did—bouncing light back and forth between two mirrors in a cavity thousands of wavelengths long. One of the mirrors would let coherent light of just the right frequency out.[12] Townes and Schawlow published their ideas in a December 1958 *Physical Review* paper, "Infrared and Optical Masers." But neither Townes nor anyone else in academia would invent a working laser, though not for lack of trying. That distinction went to a young scientist at a defense contractor a continent away from New York City.

Theodore Maiman, fresh from his Stanford physics PhD, had started at Hughes Aircraft Corporation's atomic physics department in Culver City, California, in early 1956. His first project, which would become a three-year affair in service of the US Army Signal Corps, involved a ruby maser that he and colleagues ultimately shrank down from a two-ton-plus behemoth to a four-pound mite.[13] Maiman leapt into the laser fray when that project wrapped up in August 1959. Where others looked at using gases such as potassium, uranium, samarium, or a helium-neon mix, Maiman focused on solid, synthetic pink ruby (or, technically, chromium-doped aluminum oxide, a decidedly less fetching name for the gemstone). Townes hadn't been alone in dismissing ruby as better suited for jewelry than lasers. But Maiman, steeped in ruby from his maser miniaturization work, recognized that not only could it potentially emit a bright beam, but it would also be more rugged, handle a wider range of temperatures, and enable smaller devices, such as ones that would eventually ride on the aircraft or missiles of the sort Hughes produced.[14]

Maiman figured he could use xenon flash lamps, the photographer's mainstay, for his laser pump—the whip used to jolt the ruby's chromium ions into a stimulated state primed for lasing. He paged through a General Electric catalog for the brightest ones he could find. They happened to be helical curlicues though which he could insert a ruby rod the size of his pinkie fingertip, silvered on the ends. Protruding from its stainless steel holder, it looked like an over-extended lipstick.[15] The whole thing was small enough to hold in one hand. By May 16, 1960, Maiman had run the numbers and built his experiment. He cranked up the voltage and pressed a button, and the flash fired as if he were taking a photo. As he had predicted, his creation spat out a red dot of deep-red, coherent light. Hughes held a press conference to announce the discovery of the world's first laser two months later.[16] The laser was, ultimately, a product of the most basic science there is—science that happens in the brains of geniuses long before geniuses with tools finally prove them right.

Hughes set out in pursuit of ways to capitalize on all this brilliance. The timing was good for the company. Having specialized narrowly in missiles and radar fire-control systems for air defense interceptor aircraft, it was looking to diversify.[17] Among the ideas that emerged: laser communication systems, optical radars, micromachining tools, beam weapons, submarine detectors, inertial guidance, and medical tools. Hughes settled on optical radar as most promising. It could leverage Maiman's invention as well as the microwave radar work being done at Hughes. It was an easy sell: the shorter the wavelength, the sharper the resolution. It had been widely understood that should a laser-based radar become a reality, one could locate an enemy to within a foot from fifty thousand feet away—about one thousand times better than radar could manage. Laser radar would also have much smaller antennas, an advantage for mobile systems on everything from tanks to spacecraft.[18]

The Hughes team came up with the name Colidar, for "coherent light detection and ranging," (a nod to radar's "radio detection and ranging") and built a working prototype by January 1961. It used an amateur eight-inch reflecting telescope as a receiver and a repurposed infrared star tracker as a transmitter. For the laser, Hughes bought synthetic rubies from an agent whose typical clientele incorporated them into the gears of Swiss watches. The agent arrived "with his sample case like a peddler of pots and pans," one Hughes alum recalled. "We would look at a distant object through a proffered ruby and if the image was recognizable we bought the ruby on the spot."[19]

Among other ways, they tested the device by targeting the tower of the Mormon temple and its neighboring high rises along Wilshire Boulevard in Westwood and got it to where they could nail the distance of a dark, ten-foot-wide object from about six miles away. By late 1962, they had built a more compact range finder they called the Mark II Colidar. This first-ever commercial lidar incorporated fellow Hughes researcher Robert Hellwarth's idea of Q-switching—a way to create far shorter (and more potent) laser pulses that emerged much more cleanly and predictably than those of the first-generation lasers.

**Hughes Aircraft Company engineer Rod Smith aims the Colidar Mark II, the world's first lidar. Hughes called it an "'invisible tape measure' to pinpoint remote targets such as tanks." (© HRL Laboratories, LLC.)**

The team packaged the laser transmitter and receiver into an optical twist on a double-barreled shotgun and put the electronics in a Boy Scout backpack. They made the rounds, demonstrating the device in the United States and Europe, and landed a few one-off deals for experimental prototypes, mostly for use in aiming tank cannons. When the army started putting them on their Cold War–mainstay M60 tanks as part of a Hughes-built fire-control system, the company had a new business, and lidar was a commercial product that would find its way onto thousands of war machines.[20] It wouldn't be the last military use of lidar. Meanwhile, others had much more distant targets in mind.

Louis Smullin, who ran the Active Plasma Systems group in MIT's Research Laboratory for Electronics, was among them. Born to Ukrainian immigrants in Detroit during the First World War, he was a brick of a man who had played lineman for his high school team in the days of leather helmets and was known for his fearlessness. Despite this image, Smullin was a quiet and thoughtful person who, despite a lack of interest in small talk, was happy to converse about science, teaching, politics, or other topics of the day and had a rare ability to boil complicated realities into simple descriptions. In committee meetings, Smullin was the sort who mostly listened but would invariably make a comment that changed the entire direction of the discussion.[21]

His career, like that of so many other scientists, was shaped by the Second World War. He had earned his master's degree in electrical engineering at MIT in 1939 and had hoped to go on for his PhD but was rebuffed—perhaps because he stuttered, and perhaps because with his stutter he might not make for much of a lecturer. That MIT's Electrical Engineering and Computer Science department today bestows an annual Louis D. Smullin Excellence in Teaching Award serves as yet another reminder that you can't judge an audiobook by its narrator.[22]

And so in 1939 Smullin took his MIT master's degree to Fort Wayne, Indiana, to design and test photomultiplier tubes for the Farnsworth Television Company, followed by a stint with Bendix Aviation in Sidney, New York. With war looming, he was back in Cambridge at the MIT Radiation Laboratory in 1941, where he worked on microwave transmitters and receivers for airborne radar until there was peace again. With the shuttering of the Rad Lab, Smullin spent about a year running the microwave tube group at the Federal Telecommunications Laboratory in Nutley, New Jersey, before returning to MIT in 1947, helping set up MIT Lincoln Laboratory—a Cold War–triggered reincarnation of the MIT Radiation Laboratory founded in 1952—and becoming its Radar and Weapons Division chief. He returned to MIT for good in 1955.[23]

By 1961, Smullin was forty-five years old with four kids, applying his fearlessness in his research lab, to which he rode his bike the five miles to MIT from Watertown. One fall day after parking his bike, he got a call from the MIT Research Laboratory of Electronics office about a potential postdoctoral researcher, an Italian who had been working in Canada and in New York.

"As it happened, I had recently convinced the laboratory director that it was possible to shine a laser at the moon, and to detect the reflection," Smullin recalled. As it also happened, MIT Lincoln Laboratory was finishing up a forty-eight-inch telescope for the US Defense Department's Project Defender, a missile defense research effort. It was to be delivered by the following June. Until then, a telescope so big its operator rode on a seat attached to its flank was open for business even as it was being completed. In addition, Raytheon, based in nearby Waltham, had agreed to let Smullin use what was at the time the most powerful ruby laser on the planet. The US Army Signal Corps, the Air Force Office of Scientific Research, and the Office of Naval Research would foot the bill.[24] Smullin needed help putting this all together, and Giorgio Fiocco could provide it.[25]

Fiocco was thirty-one, with swept-back wavy hair and a thick accent. Smullin described him as a "smiling, bouncy guy." Like Smullin, Fiocco had worked in radar—at Marconi in England (it was, despite its name being as Italian as Fiocco's, a British firm) and then at Cornell Aeronautical Laboratory in Buffalo, New York. A PhD electrical engineer with equal strengths in theoretical work and crafting instruments, his name was already on several patents. But his interest was less in the objects that radar found than the supposed "noise" coming from the air on the way to and from those objects.[26]

Smullin named his proposed moon shot Project Luna See, as fine a bit of wordplay as exists in the annals of science. With Fiocco doing the heavy lifting while Smullin split time among his various endeavors, they worked through the winter and spring at MIT in Cambridge. The key elements were the big Raytheon laser and a photodetector in the form of a photomultiplier tube not so different from the ones back at the Farnsworth Television Company. The photomultiplier would tune into whispers from incoming photons and shout their presence onward in the language of electrons, communicating with an oscilloscope of the sort used in the weapons-control radar systems Fiocco and Smullin knew so well. They would, during the experiment, record the oscilloscope's readings by taking photos of it. Complicating matters somewhat was the need to keep the photodetector chilled in a liquid nitrogen jacket and then sealed in a vacuum chamber.[27]

As far as the experiment's timing, they knew the receiving telescope's roughly fifty-mile-diameter purview of the moon would have to be on a dark patch of it. But the dark part of the moon varies in darkness by a factor of fifteen over the course of a month. The almost-full moon's darkness is darkest, but light leaking over from the moon's sunlit surface would interfere. They decided to go with a half-moon.[28]

In April 1962, Fiocco and Smullin spent a few days integrating their equipment at Lincoln Laboratories' Annex 4 in Lexington, Massachusetts.[29] They connected the Raytheon laser to a twelve-inch telescope that hung off the forty-eight-inch telescope like a lamprey. The smaller telescope would do the shooting. They tested it, among other ways, by burning millimeter-wide holes in aluminum foil.[30]

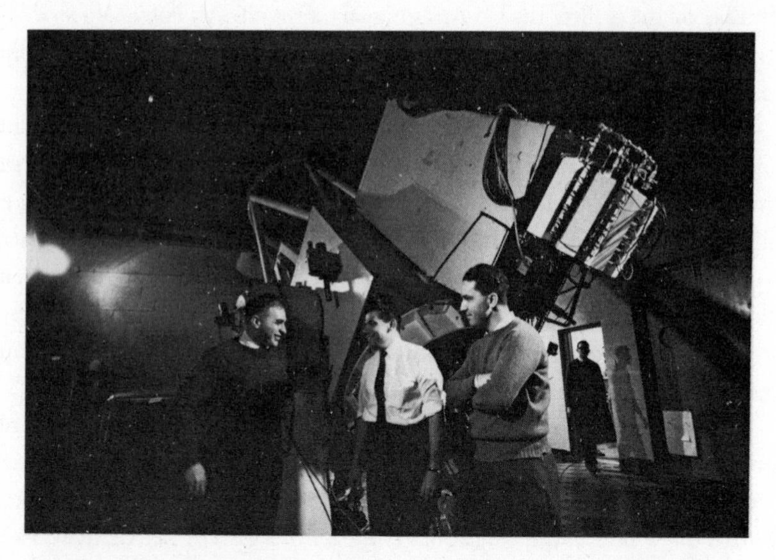

**Louis Smullin (left), Stanley Kass of Raytheon, and Giorgio Fiocco discuss Project Luna See as the telescope takes aim at the moon. (Photo courtesy of MIT Museum.)**

The laser itself was more or less of the same design as the one Maiman built, though not even close to something one could hold in one's hand. Four xenon flashlamps would blast about two million watts at a six-inch ruby rod for a thousandth of a second. The rod would then spit out a hundred-thousand-watt pulse lasting half as long. They opened the doors above the telescope to find the sky full of fat snowflakes. They waited two weeks for the next half-moon.[31]

Starting just before 10:00 p.m. on the crystal clear night of May 9, 1962, they aimed their borrowed telescope out over the treetops toward the moon crater Albategnius and fired.

Had time stopped in the split second after the first pulse, a swath of a quintillion red photons would have been visible stretching a hundred miles toward Luna. By the time the laser pulse reached Albategnius just over a second later, the light had fattened to a roughly two-mile-diameter circle. As the *New York Times* reporter on hand put it, "Had someone been standing in the mountainous region southeast of the crater Albategnius on the moon, the stark and darkened lunar landscape would have been lighted by a succession of dim red flashes."[32]

A second or so after each pulse, the photodetector back in Lexington looked for light with a wavelength of exactly 694.3 nanometers, the same deep-red color as Maiman's ruby laser and all ruby lasers hence—it's just what chromium doles out in this universe. After each of the eleven pulses Fiocco and Smullin sent to the heavens, the detector spotted about a dozen photons that, after a roughly half-million-mile swim through space, had returned home like salmon. The experiment that night took all of eleven minutes, stretching that long because the laser had to cool off for a solid minute between pulses. They followed up the next two nights, firing twenty-three shots toward Copernicus, thirty at Tycho, and another sixteen at Longomontanus.[33]

The maser-laser pioneer Charles Townes, now MIT's provost, called Luna See a "benchmark" and noted that this was the first time a laser had been aimed at a target beyond the Earth.[34] Friends of Smullin's daughter Susan, then a student at Brandeis University, took to referring to her father as "moonbeam Smullin."[35]

Her dad and Fiocco "were having fun—serious fun," she recalls. "They didn't know whether they could do it or not, and it turned out that they could, and that it was a big deal."

But she also asked him: besides having fun, why? He said it was about proving the concepts.

"This is just the beginning," Smullin told her.

His and Fiocco's next act would have nothing to do with the moon. That *New York Times* reporter noted, toward the end of his story about the night of May 9, 1962: "Although it was an exceptionally clear night, the experimenters could see a thread of red light high overhead, aimed at the moon. Apparently, there was a haze layer several thousand feet overhead."

The implications of this weren't lost on Fiocco and Smullin—any more than they would have been on Hutchie Synge. Their laser was producing the same atmospheric scattering that Edward Hulburt's and Ellis Johnson's searchlights did in the late 1930s. But the laser could do much, much more.

# CHAPTER 3
# INTO THIN AIR

Project Luna See wouldn't be the last time light from a lidar touched the moon. But the blushing of the atmosphere above Lexington, Massachusetts, would set a major course for lidar as a scientific tool—the same one established all those years ago in a quiet room in Dundrum.

Fiocco and Smullin's next experiment would be the first time a lidar was used to do atmospheric science.[1] Their big telescope moved on to NASA's facility in Wallops Island, Virginia. But the space agency and the research arms of the army, navy, and air force chipped in for a follow-on experiment. This used a downright tiny telescope—its aperture was less than three inches across, well into hobbyist territory—to send pulses from a new ruby laser built by electronics company RCA. The receiver telescope's foot-wide opening was a quarter of the size of the one Luna See used. They could downsize because lasers had gotten a whole lot better, fast.

By 1963, perhaps thirty US companies were selling lasers, and as many as five hundred were doing laser-related research.[2] RCA's laser pulses were ten thousand times shorter than those of Luna See—about fifty billionths of a second each. Because it produced less heat, Fiocco and Smullin could fire it about every fifteen seconds, four times faster than they did at the moon. And here's the big deal: despite pumping one hundred times less energy into the ruby, the RCA's individual pulses were one hundred times more powerful. The shorter the pulse (the duration is called "pulse width"), the more oomph a given amount of input power gets you, and the better the chance that some of those photons make it back to your detector.[3]

Shorter pulses meant a much-improved ability to use lidar to measure distance. With Luna See, the hundred-mile phalanx of red photons marching moonward meant that even if Fiocco and Smullin could have precisely measured the timing of a returning photon, they still wouldn't have known where that packet of light had been traveling within the hundred-mile band. That's why their experiment didn't bother measuring the distance to the moon itself—traditional approaches were more accurate. And they could only guess how high the atmospheric scattering was coming from. But now, with this RCA laser, they could theoretically know how far a photon traveled to within about fifty feet, assuming the electronics were fast enough to track with absolute certainty the timing of when a pulse was sent and received.

While good receivers had arisen from two decades of radar development, they weren't perfect. Still, with the RCA laser, they could easily divide the atmosphere into ten-kilometer increments and, knowing that light travels about thirty centimeters (one foot) in a nanosecond, keep rough tabs on photons received from different altitudes by telling the receiver's photodetector to open its nitrogen-cooled eye at slightly different times. And that's what they did.

Over nine nights in the summer of 1963, they pointed their hardware straight up into the graveyard-shift skies over Lexington. Their quarry was light reflecting back from various altitudes, which they recorded by taking photos of an oscilloscope after each of two hundred to four hundred laser pulses. They got clear returns all the way up to nearly 130 kilometers (81 miles). At heights above 30 kilometers (19 miles), they found, "signals were so weak that it was possible to count individual photoelectrons," which made sense: the stratosphere tops out at 30 kilometers, at which point 99 percent of the atmosphere is below. They speculated that echoes from 80 kilometers up were from noctilucent clouds and that those at 120 "correspond to the region of meteoric break-up."[4]

It was the last lidar experiment Smullin would do. He turned his attention to high-power radar systems and microwave amplifiers, electron beams, and plasma dynamics—in addition to becoming a department chair at the place that had rejected him as a PhD student. Fiocco stayed with lidar, shooting the skies above Massachusetts, Alaska, Greenland, Antarctica, and many points between in the decades to follow.

Fiocco and Smullin weren't the only ones aiming lidar skyward in mid-1963. Myron G. H. Ligda, known as Herb, was leading the Stanford Research Institute's (SRI) Aerophysics Laboratory at the time. The nonprofit Stanford University research affiliate in Menlo Park, California, was about thirty-five miles from where he grew up in Oakland, but Ligda had taken a roundabout route getting there. He graduated from high school in 1936 and joined the marines as a French horn player rather than as the grunt he aspired to be, his poor eyesight ending dreams of amphibious landings. He left; enrolled in the University of California, Berkeley; marched in the band; was a diver on the swim team; and, before graduating, took a job with the US Weather Bureau as an assistant observer at Oakland Airport Station. When he graduated, the Weather Bureau transferred him to Washington, DC, where he did additional training before enlisting in the army in 1942 as a weather observer specialist.[5]

Ligda was officer material and climbed the army's weather-forecasting ladder, landing in advanced training at New York University and then training at Fort Monmouth in New Jersey "on equipment of a nature I can't disclose," he wrote. This was radar, and he ended up in the army's Sixth Weather Squadron in Panama in early 1944.

Weather radar was a new thing—pioneered in early 1941 when researchers on the south coast of England pointed a ten-centimeter air defense radar at a hailstorm and tracked it for seven miles.[6] In Panama, Lieutenant Ligda managed the first-ever network of weather radars. With two land-based and two airborne radars, Ligda and colleagues tracked storms and studied how winds and the land-sea boundary affected storm behavior. His Panama team was the first to identify radar echoes from lightning, as well as from migrating birds.[7]

Ligda left the service for MIT, where the US Army Signal Corps was pouring money into weather radar research after the war—enough that in addition to two research radars on the roof, Ligda and colleagues had a B-17 Flying Fortress bomber at their disposal to fly instruments into the storms their radars probed.[8] Ligda had earned his doctorate in meteorology by 1953 and took it to the Texas A&M in College Station, where on April 5, 1954, his team spotted a tornado heading toward nearby Bryan, Texas. They got the word out with twenty minutes to spare, with police calling schools and two radio stations urging folks to take cover. Some two hundred houses were damaged, but nobody was hurt, besides "a slightly injured dachshund" and Ligda himself, who came home with a gash in his forehead sustained during one of his forays outside to check out the approaching fury.[9]

As much as he liked the work, Ligda wasn't into Texas: "The heat, insects, lack of availability of metropolitan life, and rather narrow circle of acquaintances all bother me somewhat," he wrote. Racial segregation bothered him too.[10] In 1958, he headed back to California, to SRI. His focus was continuing the work he had started in Panama and continued at MIT and Texas A&M, which was to combine many radars over thousands of square miles for better forecasting of hurricanes or the thunderstorm systems of the tornado belt. But he kept an eye on technological developments and was interested in the optical version of radar some were calling lidar. Ligda had seen radar go from a way to spot enemy aircraft to a tool capable of tracking tornadoes in the span of a few years. He wondered if an optical version of radar could become a similarly valuable tool in meteorology. He happened to be in the right place to find out.

Workers in SRI's converted Dibble Army Hospital building, where Ligda worked, enjoyed scheduled morning and afternoon coffee breaks. This may seem quaint, but it was strategic. SRI had well over a thousand employees at the time, working in computing, optics, telecommunications, medical technology, materials science, and a lot more. Coffee breaks were a way to get outside one's silo. During one of them, Ligda chatted with an Electromagnetic Techniques Lab guy named Dick Honey. He wondered if Honey's group could build a laser that Ligda could make into a tool to study the weather. Sure, Honey said.[11]

Ligda had no fear of the technical unknown. In high school, he cobbled together his first car out of junkyard parts; after the marines, he built a radio into a diving helmet and, together with some pals, did underwater salvage work and boat repairs. While at MIT, he built his family a two-bedroom house in the woods of Middlesex County, Massachusetts, despite having zero construction experience.[12] This was more of the same. By July 1963, Ligda's and Honey's teams had mashed up an industry-donated ruby laser, a photodetector, and some government surplus optics on a whitewashed antiaircraft gun mount they bolted to the roof of the building. They called it the SRI Mark I lidar. When they shot it into the sky, they aimed lower than Fiocco and Smullin had, at where weather happens. It was the first time anybody had done that, and they quickly showed lidar's value in spotting cloud, smog, and haze layers; temperature inversions (in which it's colder lower and warmer higher, trapping the cooler air and, until a strong wind clears it out, letting air pollution fester); cloud structure; and ceiling height.[13]

The SRI lidar worked much like the Hughes Colidar did—it aimed laser light at a target and figured out how far away it was based on the time it took a tiny fraction of the light to bounce back. The targets were just smaller here: particles in the air—fog, dust, pollen, steam, haze, smoke, smog, the clouds above—collectively called aerosols. Radar waves, being much longer, were better for seeing bigger things like raindrops and snowflakes, but radar ignored aerosols (and still does). This was the big opportunity for meteorological lidar, and Ligda; his deputy, Ron Collis; and an eight-person SRI team including engineers, scientists, optics experts, electronics experts, and a machinist chased it.[14]

**Charles Northend (left), Herb Ligda, and Ron Collis on the roof with their SRI Mark II lidar. (Photo courtesy of SRI International.)**

Collis had joined the Royal Air Force at age nineteen in 1940 and was a pilot during the Second World War, later graduating from Oxford and serving in the Royal Navy as a meteorologist.[15] As with about everyone else involved in the early days of lidar, he was a radar expert. Collis was at least as enthusiastic about this optical radar as Ligda and published prolifically about the technology's development. Among those reading his and his colleagues' publications was a young meteorologist named Warren Johnson.

Johnson had been a US Air Force ROTC student at Texas A&M whose military aims were, like Ligda's, thwarted by poor vision (Johnson had hoped to be a pilot). Soon after graduation, the air force had sent him a questionnaire with a dozen or so nonflying job choices. Being a newly minted electrical engineer, he checked three engineering boxes. He was surprised when he learned that he was being sent to meteorology school at the University of Wisconsin. His upbringing in a West Texas ranch town always needing rain had instilled a latent interest in weather, so he wasn't unhappy with this news.

The air force had needed more weather forecasters, so they pulled in the likes of Johnson and put them through a yearlong program in Madison and elsewhere. A couple of years later, when he had fulfilled his obligations to the military and was working as an engineer in nuclear ordinance at Sandia Corporation in Albuquerque, Johnson realized he missed meteorology. He got a Ford Foundation grant and returned to the University of Wisconsin in September 1961. Legends like the atmospheric turbulence guru Heinz Lettau and Verner Suomi, the father of weather satellites, were on the faculty. Johnson's research focus was the dynamics of the lower atmosphere—the first mile or so of air. To find out how much frictional energy was being sapped from the lower atmosphere by the "rough" forests of northern Wisconsin, he and his team measured vertical wind profiles throughout the year by releasing and tracking hundreds of small balloons. In the winter, he recalls, "it was way too cold to be much fun."[16]

With his PhD in hand in 1965, Johnson got a job with the National Oceanic and Atmospheric Administration (NOAA) in Oak Ridge, Tennessee, where the massive Oak Ridge Gaseous Diffusion Plant had enriched uranium since the Manhattan Project days. NOAA's job there was to understand how the rough terrain surrounding the plant would influence the dispersion of nuclear material and what the doses might be in case of nuclear disaster. While working here, he came across Ligda, Collis, and company's work using lidar to probe the lower atmosphere.

"I thought, wow, they're using the aerosols to determine what the structure of the atmosphere is, and basically they could see stratifications caused by temperature inversions that are going on," Johnson said. It sure beat sending a balloon up during the winter in northern Wisconsin.

Johnson interviewed with Herb Ligda at SRI in 1967, and Ligda hired him. Johnson's arrival was, in a way, emblematic of how lidar was evolving as a tool of science. Ligda and Collis and others in the first generation of scientists were meteorologists, true. But they came from the world of microwaves and radar, just as the originators of the laser itself had. Johnson wasn't a radar guy at all, and while he was an electrical engineer, lidar technology had come along to the point that his expertise applying lidar in answering scientific questions was enough.

Hiring Johnson turned out to be one of the last things Ligda did at SRI. He was already wracked with colon cancer; one day soon after Johnson started work, Ligda "went to the hospital and never came back," as Johnson recalls. He died that October at age forty-seven.

By the time Johnson arrived, SRI had built five generations of lidar. As contract researchers, they took work from all directions, the only condition being that the results would fall where the science took them, regardless of where the client would have liked them to lead. Once they had an idea how various lidars depicted various things happening in the sky, they got more specific: figuring out if the technology could be used to assess cloud cover or fog for aircraft coming in for landings; using their first airborne lidar to track the dust cloud after an explosion near Fort Peck Reservoir, Montana[17]; and, increasingly, looking at plumes, which was what Johnson did on his first project.

With Collis and a couple of technicians, he ended up back in the forest, this time in Idaho's Sawtooth Mountains with SRI's Mark V lidar instead of a flotilla of balloons. There had been a beetle infestation, so the US Forest Service was spraying insecticide via helicopter. They had contracted SRI to figure out where it would drift.[18] A year later, in 1968, Johnson and colleague Ed Uthe headed to Pennsylvania with the same lidar mounted in a truck; by then, it could fire pulses every five seconds, though they were still recording data using a Polaroid camera and then interpreting the photos into punch-card input in what was euphemistically called "hand digitization." The goal was to understand how well lidar could establish where the plume pouring out of the new Keystone Generation Station's eight-hundred-foot smokestack was going.[19] The next year, Johnson and Uthe had their Mark V lidar in an air force C-130 zigzagging over a roughly hundred-thousand-square-mile patch of ocean east of Barbados, aiming the device generally downward out the jump door as they harnessed in to avoid being sucked out into the 225-mile-per-hour slipstream. This was part of BOMEX, a three-month summer campaign involving seven federal agencies' ships, buoys, aircraft, balloons, and satellites all aiming to help figure out how Atlantic hurricanes get cranking.[20] Johnson and Uthe's lidar was the first to spot what Johnson called "a really strong dust layer" floating on a raft of cooler air (an inversion) a mile or so over the Caribbean Sea. This is now called the Saharan Air Layer, and its role in the tendencies of Atlantic hurricanes remains a subject of research.[21]

There was much more in the coming years: they mapped smog layers and looked at regional air quality in places like St. Louis and Los Angeles, and they developed lidar tuned to track chemical warfare agents.[22] And they were far from alone, as groups around the world embraced lidar as a way to see the previously invisible action happening nearly everywhere in the air around us.

The atmospheric lidar community used a growing selection of lasers to exploit the fact that different aerosols reflect laser light of different wavelengths—which meant that it was possible to not only see stuff floating around but identify what it was. Shorter laser wavelengths could spot smaller and smaller particles—all the way down to specific molecules as time went on. By using two lasers at once, each tuned to a slightly different color, you could tell how much of a particular gas or aerosol was out there. Doing this with water molecules, for example, would have obvious implications in weather prediction.[23] This technique is known as differential absorption lidar, or DIAL. Richard Schotland, a PhD-student colleague of Ligda's at MIT, went on to pioneer the technique at New York University. It may hold the key to better severe weather prediction, as we'll see.

There's one more early development in atmospheric lidar technology to note, and that is the advent of coherent lidar. This branch of lidar technology quietly emerged a couple of years after Fiocco and Smullin's moon shot. Then, as lasers, computers, software, and all the rest moved ahead, it flowed into the mainstream. It's not to be confused with coherent light—light where the peaks and troughs of the light waves align more or less perfectly, which is the defining feature of all lasers.

All the lidar we've talked about so far has involved bouncing photons out to some target—the moon, an enemy tank, Saharan dust floating a mile or so above the waters east of Barbados. These lidars treat returning photons like little tennis balls thumping against the garage doors of their photodetectors. They can't tell what color the balls are; they only know that they've been hit at a particular time. Another term for this is *direct detection* ("incoherent detection" works too, though people who dedicate their careers to it can be sensitive about that). Getting back to the particle/wave nature of light, you could say these systems have all focused on the particles. *Coherent detection*, in contrast, pays attention to the full wave nature of light coming back. It wants to know all about the tennis ball. This is harder to do, and that's why it took longer to get going. "The technological and practical difficulties are considerable," SRI's Collis wrote in 1970, "and the range at which such measurements may be made appears likely to be limited for some time."[24]

Forty-seven years after Collis wrote those words, I'm seated across the desk from the guy who had, as of 1970, already done a lot to overcome those practical difficulties.

# CHAPTER 4

# PLANE CRASHES, HAILSTORMS, AND DISTANT PLANETS

In this office building in Broomfield, Colorado, between Denver and Boulder, Milton Huffaker looks different than in the photos I've seen of him, which were taken at NASA's Marshall Space Flight Center in the mid-1960s. That was just a couple of years after Huffaker had left his PhD studies in low-energy nuclear physics at the University of Kentucky to work on the Saturn rocket program. Back then, he had the standard-issue crew cut and a tie like a typewriter ribbon. His hair is longer and gray now, and he wears a turquoise fleece over his blue Oxford despite it being August. The knot of his bolo tie is a silver cardinal. He explains, slowly and precisely, how he went from there to here, pioneering wind-detection lidar and saving untold lives along the way.

In Alabama, Huffaker was involved in studying what an Apollo space capsule would experience when flashing through the atmosphere on the way back to Earth, and then on understanding the hellish environment created by the exhaust plumes from the Saturn's second stage. As the rocket's design firmed up, NASA management wondered whether it might be possible to measure the winds the rocket would fly through on the way to space—and then, more broadly, whether lasers might enable more precise wind detection in general. That would take coherent detection, which could determine the motion of air based on the Doppler effect. The Doppler effect is why a passing car sounds higher-pitched on approach than after it passes. The speed of sound is constant at a given temperature and air pressure (about 761 miles per hour at sea level on a pleasant day). Incoming speed squeezes the sound's waves as if pressing down a spring. This pushes the pitch we hear higher. When the car has zipped by, the opposite happens.

The Doppler effect works with light too. It's how we know the universe is expanding: light from distant galaxies hits our telescopes with longer wavelengths than when the stars emitted it billions of years ago, a phenomenon called redshift. Coherent detection lidar exploits this. The idea is to compare the frequency—the optical "tone"—of light waves that have bounced off some target with the steady frequency of the original laser light.[1] If the air and the stuff floating in it is moving toward the laser, the returning frequency will be higher than the one sent; if the target is moving away, the frequency will be lower.

I belabor these sorts of details because Huffaker ultimately built a company, Coherent Technologies (subsequently bought by Lockheed Martin), that makes coherent lidar to prevent plane crashes at airports all over the world. The Japanese and others are now installing their own systems on aircraft that can help avoid the clear-air turbulence that sends flight attendants, drinks, and unbelted passengers ping-ponging about airline cabins. The European Space Agency is spending around $500 million on a satellite called Aeolus that uses coherent lidar to understand global wind patterns at elevations up to thirty kilometers (nineteen miles), which promises to vastly improve weather forecasting (it could also help optimize airline routing and boost the accuracy of long-range weapons). But we're getting ahead of ourselves.

To make a coherent detection lidar, you need an extremely stable laser, since it's the baseline against which you're trying to measure the Doppler effect. No laser is perfect; early ruby lasers, even after Q-switching, were much too unstable. As the 1960s rolled on, though, other choices arose. Bell Labs had a helium-neon gas laser working before the end of 1960—one that emitted a continuous beam of the sort we all imagine lasers making. Teams at General Electric and IBM invented different semiconductor lasers within two weeks of each other in 1962.[2] Bell Labs announced the carbon dioxide laser in 1964, the same year Hughes invented the argon-ion laser. Soon enough there were lasers made from the vapors of copper and other metals; liquid lasers, dye lasers, and organic lasers; and lasers, similar to ruby, made of crystals such as calcium fluoride and yttrium aluminum garnet (YAG).[3] There were tons of lasers, in short, and they just kept coming.

Huffaker worked with helium-neon and argon-ion lasers but ultimately settled on a carbon dioxide laser—a continuous-wave laser that shot out a constant beam in the infrared, with a wavelength about fifteen times longer than that of the ruby laser.

"$CO_2$ lasers were hard to build; their detectors were helium cooled," Huffaker says. "It's a difficult laser to be working with."

But it had advantages: the longer the wavelength, the easier it is to make it stable. And the $CO_2$ laser was stable enough that, with a lot of work, it could be used as a basis for detecting the movement of a breeze without having to stick a glorified pinwheel into it. There was precedent: in 1963, a Columbia Radiation Laboratory team had shown that a coherent lidar setup could spot the motions of microscopic plastic spheres in a liquid.[4] Huffaker was looking for microscopic particles in air—fast-moving air, at supersonic speeds in small wind tunnels, since aerodynamics was his focus. The Federal Aviation Administration got interested: Could a laser help detect wind shear? How about aircraft wake vortices? Both were invisible killers. Wind shear, a product of microbursts associated with thunderstorms, pulls the rug out from under planes, aerodynamically speaking, on takeoff and landing. For years, it was the leading cause of airline accidents.[5] A wake vortex is to an aircraft what a wake is to a boat, but it comes in the form of invisible tornadoes at either wingtip of large aircraft. These tornadoes can crank at two hundred miles per hour or more, causing smaller planes to lose control. Because of wake vortexes, the FAA mandates gaps between takeoffs and landings of aircraft that differ depending on how big the planes in question are. Huffaker had firsthand experience with a wake vortex.

"I got behind a DC-6 in a Piper Tri-Pacer. I got flipped to the right—didn't do a 360, I got out," Huffaker says. "Anytime you find yourself in that situation, you're lucky not to be upside down."

Huffaker's work expanded into the possibility of using coherent lidar in aircraft themselves to spot clear-air turbulence in the skies ahead. That's only now coming to fruition, at Boeing and elsewhere.[6]

When NASA lost interest, Huffaker continued his work at the NOAA in Boulder, where he helped build coherent lidars capable of seeing the atmosphere in new ways. In 1984, at his Coherent Technologies startup in nearby Louisville, Colorado, continuous-wave gas lasers gave way to pulsing solid-state lasers, and then to the combination of dye and solid-state lasers used in the Coherent Technologies WindTracer product Lockheed Martin now sells to airports and wind farms around the world. It scans the skies, pulsing 750 times a second, looking for Doppler shifts from the aerosols in winds as far as twenty miles away. On the wake-vortex side, WindTracer has helped the FAA better understand the gaps between successive takeoffs or landings necessary at busy airports. By shortening those intervals based on Doppler lidar data, Delta Airlines has saved tens of millions of dollars in annual fuel costs at Atlanta's Hartsfield-Jackson airfield alone while cutting runway departure lines by around 20 percent. At FedEx's Memphis hub, tighter spacing has increased the number of hourly takeoffs and landings by about 20 percent and saved $1.8 million a month in fuel burned by idling jets.[7]

Coherent lidar has become a field of its own, with innovations coming from Europe, Asia, Australia, and the Americas. Military leaders are interested in it to enable more accurate air drops; wind energy companies use it to optimize how they feather the blades of their enormous turbines; atmospheric scientists use it to study ozone pollution and how the layers of the atmosphere interact. It can be used to identify a distant vehicle based on its vibrational signature.[8] And as mentioned, coherent lidar is headed to space: it's the sole scientific instrument on the European Space Agency's Aeolus mission to measure global winds, slated for launch in 2018.

Still, the vast majority of modern lidar, atmospheric and otherwise, is the tennis-ball-against-the-garage-door direct-detection/incoherent variety. The progress made in the decades since Fiocco and Smullin took aim at the moon, in terms of lidar technology as well as the diversification of its uses, is epic, and to cover it all respectably would take a shelf full of books like this. It's used in studying the climate, conducting air quality and pollution studies, sensing trace gases both natural and human-made (think detecting chemical attacks),[9] finding leaks from natural gas pipes, understanding the innards of clouds and the impact of volcanic eruptions on planet-cooling stratospheric aerosols, and a lot more. So I went with two examples to give you an idea of how far things have come. The first promises to make a big difference in weather prediction, in particular as relates to wicked storms. The second could help us better understand the nature of distant planets as well as our own.

The story of lidar for better weather prediction starts with one of eight kids who grew up on a small dairy farm in rural Wisconsin. In the depths of the Great Depression, Ed Eloranta's father had to leave school in the eighth grade to work on the family farm. His mom made it through high school.

Perhaps because they hadn't been beneficiaries of it, they placed enough value on education to invest an Encyclopedia Britannica set for the family. Ed, the oldest, devoured it, particularly the scientific stuff, using its teachings and those of borrowed books to build his own rockets and grind the mirrors for his own telescope. Although Eloranta was a good student, his high school counselor tried to dissuade him from attending the state university in Madison "because it was too hard. Everybody who came from here flunked out," Eloranta recalls.

He enrolled in the University of Wisconsin anyway. He had $400 from his folks to support his education—all they could afford with seven kids behind him. Even in 1961, this wasn't going to cover it. So he worked. He also studied physics, which his college counselor advised against—more so when he struggled, having come in so far behind. "But I soldiered on," Eloranta says.

When he was a junior working on the side in a high-energy physics lab, he was faced with the prospect of having to solder fittings to the ends of three thousand hair-fine cables destined for a high-energy physics experiment. He happened to be taking a course on the physics of the upper atmosphere (the ionosphere's influence on radio signals interested him). The professor, fresh from his PhD, was Stig Rossby, son of the man after whom Rossby waves—sweeping undulations in the jet stream with huge influences on regional weather—were named. Eloranta asked Rossby if he needed anyone to work on electronics. Rossby did. "It paid a nickel more an hour, and there were no cables," Eloranta says. The following year, 1965, Eloranta started a PhD under Rossby. "Suddenly, I became a meteorologist rather than a physicist."

For a year or so, he punctuated lab work with flights in NASA aircraft over the Caribbean and field programs in New Mexico, the ultimate aim being to develop a satellite experiment for lightning detection. But that ended when Rossby didn't get tenure.

It was 1966, and Eloranta was in the unenviable position of being a PhD student without a professor. He settled on conducting an experiment having to do with electrification and fog. A PhD student, John Regan, had built a ruby lidar for another experiment. Eloranta asked him to bring the device—a six-inch-diameter telescope, a photomultiplier detector, and an oscilloscope with a camera on it to record signals—over. It fired once a minute. Eloranta used the data to come up with a theory on something called multiple scattering lidar.

Multiple scattering is simple enough in concept: most of the photons bouncing off fog, dust, or other ethereal targets come right back like a tennis ball off a racket. But some of them ricochet off other particles first and then back to the receiver. Eloranta's master's thesis, and ultimately his 1972 PhD dissertation, had to do with the math to represent this, which would in turn improve how lidar data were interpreted.

While still a student, he had written a grant to the army to measure wind speed as a function of altitude. Huffaker had done the same thing with coherent lidar; Eloranta's lidar achieved something similar by taking the equivalent of laser snapshots in quick succession and seeing how the dust moved. Though he wasn't offered a faculty job, he stuck around in an academic staff position, having tenured faculty sign his proposals as the nominal project leaders and, eventually, sign the PhD dissertations of students Eloranta mentored. His team built a succession of lidars. The farm-boy-turned-scientist was good with his hands and could build and fix things. He was, he felt, neck and neck with SRI, except "they had money and I didn't." But there were benefits to being at Wisconsin and in a distant orbit of Verner Suomi. He bought a satellite display console from Suomi's team and got a student to adapt it to display lidar data in real time, no Polaroid film needed. The grants kept coming, his team grew, and his focus became increasingly on solving a tougher problem.

Giorgio Fiocco, Smullin's old collaborator, had provided the inspiration in the form of a paper he wrote in 1971.[10] Eloranta knew him too: "He was a guy who had a personal and intellectual sparkle to him," he says. Fiocco had continued probing the atmosphere with lidar long after Luna See. Some of his work focused on trying to validate the lidar returns from the edge of space that he and Smullin had first noted in 1963; some of it focused on the stratosphere. But Fiocco had also noted a fundamental problem with any lidar trying to figure out how much aerosol is actually up there. Standard lidars could gauge that there was a little or a lot of something, but they couldn't quantify it. Quantifying it is important: the amount of dust in the stratosphere after a volcanic eruption will tell you how much things might cool off. The amount of water vapor in the air can help predict whether clouds will form and rain will come.

The problem with quantifying it was that lidar's laser light isn't in a vacuum to and from its target aerosols. It passes through air molecules and random aerosol particles on the way there and back, and some of the light gets scattered by them. So while a standard lidar might be able to tell you there's aerosol up there and even where it is, you can't know how much is there unless you account for the molecular scattering.

Fiocco came up with one approach for that 1971 paper; Eloranta and the professor he worked with, Jim Weinman, outlined an alternative instrument design based on Fiocco's work. Both involved using filters to separate the molecular scattering from the aerosol scattering. They and others put together a proposal to NASA, which eventually funded it in 1975. The proposal said, in essence, two things. One was that understanding aerosols is important to modeling Earth's climate (very true). The second was that one of these instruments on the forthcoming space shuttle would be the way to understand aerosols better. That was, Eloranta says, "totally outrageous."

"Nobody had built one of these things. It was very complicated, we were working with primitive technology, and it was supposed to go to space," he says. "But one of the things that Weinman was good at was proposing outrageous things and eventually getting money for it."

They called it high spectral resolution lidar, or HSRL.[11] The system needed six-inch-diameter optical filters called Fabry-Pérot etalons that had to stay more or less perfectly flat and perfectly parallel (both to within a few nanometers) despite being temperature sensitive and having to take off in airplanes and eventually the space shuttle, a beastly ride. The lasers themselves—involving lasers to pump other lasers—gave one hundred pulses per second at a time when one pulse per second wasn't too shabby. They somehow got the system to work well enough for a 1978 test flight at NASA's Wallops Island facility. By 1980, they had done an actual air quality study.

Weinman moved on and suggested Eloranta do the same, having "gotten the glory and demonstrated that they could do it," as Eloranta puts it. "The only thing left was hard work." Eloranta soldiered on, for nearly four decades now. He and his team, now known as the University of Wisconsin Lidar Group, figured out how to use HSRL to tell the difference between water and ice clouds and, later, how to measure the size of cloud particles—the latter based on exploiting the same backscatter effects Eloranta had worked to disentangle as a student.

Over the years, they built a bunch of HSRLs, each generation more capable and user-friendly. They have probed above various continents, oceans, and the Arctic. The team refined the technology to the point that its Arctic HSRLs worked unattended in Barrow, Alaska, and on Ellesmere Island in the Canadian Arctic for years starting in 2005.[12] They built a low-maintenance mobile unit into a Winnebago and named it BagoHSRL, which remains their testbed for new HSRL advances. And for the National Center for Atmospheric Research (NCAR), Eloranta's team built a version designed for the National Science Foundation's Gulfstream V research jet. That one, called the GV-HSRL, was delivered in 2010. It has flown from California to Hawaii looking at cumulostratus clouds and west of Chile as part of a mission to track trace gases and aerosols over tropical oceans.[13] Mostly, it operates through a ceiling porthole from a lab on the fourth floor of NCAR's Earth Observing Laboratory in Boulder, Colorado, which is where I found it, and which is where weather prediction comes in.

The GV-HSRL is a black cannon protruding from the side of a white box that's full of electronics and about the size of a Xerox machine. The cannon swivels so it can point up or down through portholes in the top or the bottom of the Gulfstream V. Here, down gets it a close-up of the lab's tile. NCAR scientist Scott Spuler leads the two-man team presiding over the instrument's operation. Lately, they've been focused on using the lessons Eloranta embedded in this million-dollar hardware to create a much cheaper version of it that could pulse away unsupervised for months at a time at hundreds of sites across the country. That cheaper system, which they've shown to work nearly as well as the big GV-HSRL,[14] is based on off-the-shelf laser diodes—semiconductor lasers not so different than those in laser pointers and supermarket scanners. They call it DLB-HSRL (for diode laser-based high spectral resolution lidar[15]), and they would put it in a box with two other lidars. The combined package would involve three lasers, two of them split into subtly different colors, shooting five infrared beams into the sky.

Spuler is trim, clean-cut, with dark hair neatly parted on the side. On this summer day, he wears shorts and a gray-brown T-shirt emblazoned with the words "Costa Rica." He came to NCAR after a Colorado School of Mines PhD developing lasers to look for mercury pollution from coal-fired power plants, then a stint at a telecommunication startup in Boulder. That ended with the dot-com bust.

"We went from, I don't know, I think 150 people down to five in one day," Spuler says. Adding to the fun, Spuler had a nine-month-old daughter and had crutched into his layoff meeting, having just broken his leg on an icy trail run. He had never heard of NCAR but ended up on staff there in 2002. His first big project was a joint effort with a German team from the University of Hohenheim, led by Volker Wulfmeyer. The idea was to build a big, expensive lidar capable of measuring moisture in the atmosphere—WV-DIAL, for water vapor differential absorption lidar.

Asked his age, Spuler pauses before coming up with thirty-seven. This seems entirely plausible until his colleague Matt Hayman chuckles. Hayman, in shorts and a "Making the Moose out of Life" T-shirt, is closer to that age, with an electrical engineering PhD from University of Colorado. Spuler is a decade older, well-preserved through a combination of happy genetics, distance running, and sun-protected environments such as this laboratory in which his, Hayman's, and their collaborators' creations reside.

Spuler and Hayman's new five-laser system, which they're tentatively calling the Lower Tropospheric Observing System, or LOTOS, would plug one of the biggest holes in weather forecasting. It would sharpen five- to seven-day forecasts by improving the accuracy of computer models that attempt to predict weather at what those in the business call the mesoscale. The mesoscale spans big weather systems such as Western hailstorms, Great Plains tornadoes, Midwestern floods, and nor'easters. With the exception of things like regional heat waves and major hurricanes, the mesoscale covers the vast majority of the severe weather that farmers, airlines, transportation departments, electric utilities, military and homeland security leaders, insurance companies, and the rest of us need to pay attention to.

As an example, better forecasting on the mesoscale could help predict the emergence of the vaporous anvils that spawn hailstorms—hailstorms like the one that, a few weeks before I met Spuler and Hayman, bombarded Lakewood, Colorado, about twenty-five miles south of this NCAR lab, with frozen baseballs. The estimated $2.3 billion in damage from the May 9, 2017, storm included $873 million to replace smashed windshields and pocked body panels of whatever percentage of the 167,000 cars in the storm's way weren't totaled outright.[16] Improved mesoscale forecasting wouldn't have helped the hundred thousand houses that suddenly needed new roofs. But with even an hour's notice—rather than a few minutes—drivers might have moved their cars to shelter or out of the storm's path.

Mesoscale and all other weather prediction depends heavily on computer models. Computer models depend on the assumptions and initial conditions plugged into them. Without a decent understanding of what's actually happening in the atmosphere, it's garbage in, garbage out. And while chaos theoreticians may contend that the flapping of a butterfly's wings can trigger tornados, meteorologists focus on three big things when attempting to predict the weather: wind, temperature, and water vapor. Weather prediction gurus would like better data on all three, but water vapor remains by far the most nebulous of them (sorry).[17]

Measuring water vapor isn't always hard. A $20 household digital weather station can tell you the humidity. What's hard is measuring water vapor constantly, from the ground to the top of what's known as the planetary boundary layer, with a clarity that lets you map its ever-changing vertical structure at a given point in time. Meteorologists think having a better handle on that structure, and in particular on the behavior of water vapor in it, is the key to sharper severe weather predictions, Spuler says. Depending where you are and what time it is, the boundary layer tops off somewhere between a few hundred meters in altitude to five kilometers (three miles) up, though typically in the two- to three-kilometer range. Think of the boundary layer as an invisible elastic ceiling. Below it, the shape of the land and the heat radiating off it cause turbulence and disorderly winds. That motion changes the nature of water in the air, causing it to condense into droplets or freeze into ice. The energy captured or released can fuel potent vertical thrusts of air. That motion triggers most types of severe weather, which are ultimately unleashed from systems extending well above and below the boundary layer itself, says Richard "Rit" Carbone, a senior atmospheric scientist at NCAR. An ability to recognize the planetary boundary layer's triggers could hold the key to improved nowcasting and short-range weather prediction. A better handle on water vapor will be central to that.

The standard water vapor measurement approach is to tie a fancy version of a household hygrometer (which measures humidity) to a balloon, send it up with other sensors, and tune into its radio as it rises up through the top of the boundary layer. These balloon radiosondes are thoroughly modern hardware made by the likes of Lockheed Martin—they've got GPS to measure wind speed; precise hygrometers, thermometers, and barometers; they communicate in the four-hundred-megahertz meteorological band; the whole bit. But they're basically much-evolved versions of the balloon sondes the US Weather Bureau started lofting back in 1937.[18]

The problem with radiosondes isn't the quality of data but the sparseness. There are just sixty-nine radiosonde launch stations in the continental United States—one or two in each state on average, with several states, including Indiana, Kentucky, and West Virginia, having none at all.[19] Each site releases a radiosonde balloon twice a day, within an hour of midnight and noon UTC, which corresponds to four or five hours earlier on the US's East Coast, depending on daylight saving time. Just the time zone differences introduce variability—balloons launched at noon UTC in New York rise just as the morning sun heats things up and stretches the boundary layer higher, but the ones launched at noon UTC in California do so before the sun rises.

Weather forecasters have a lot more than just the sondes to work with, of course. Weather satellites fill in some of the picture, for example, and Doppler weather radar can see precipitation and how it's moving. But radar can't see water vapor until it condenses into droplets, and satellites leave gaps, either in terms of pixel size (with geostationary satellites 22,300 miles away) or spotty coverage (in the case of low Earth orbiters, which zip around the planet every ninety minutes or so).[20] The tango of water vapor and temperature up through the top of the boundary layer can change tempo and energy in minutes—and the convection it triggers can bring nasty weather. Bottom line is that lidar systems capable of measuring water vapor could enable more precise prediction of severe weather, well before storms take shape and show up on radar.[21]

With better insight into temperature and moisture up through the boundary layer, forecasters could recognize the signs of a hailstorm and provide notice before it starts dumping ice balls—much like Herb Ligda's team did for that tornado in Texas. "Hail is a great example. Thirty minutes of warning is plenty of time to put vehicles and other targets under cover, saving millions in one relatively small storm," Carbone says. "This is especially true in cities, since the heavily built environment is where the largest losses are experienced." Farmers could benefit too—for example, by getting animals to shelter, he says.[22]

This is where Spuler and Hayman come in. Spuler is a builder of laser systems that exploit the sorts of laser hardware his doomed telecom company worked to advance. The fiber-optic lines carrying our voice and data traffic are crammed with infrared laser light. The telecommunications industry's continued, massive investments in micropulse laser diodes has made those and other components of phenomenal capability and low cost available for this very different application.[23]

Lidar has been capable of doing water vapor measurements for two decades. Working with Kevin Repasky, a diode laser expert at the University of Montana, and his PhD student Amin Nehrir, Spuler and Hayman first built a diode laser version of the previously enormous and expensive moisture-sniffing WV-DIAL systems of the sort Spuler had worked on in Germany early in his NCAR career.[24] Spuler's and Hayman's work with Eloranta's GV-HSRL got them thinking about what they could do with low-cost, off-the-shelf, telecom-derived lasers at infrared wavelengths that are eye safe (unlike GV-HSRL's laser, which is green and not a great idea to stare into). They're doing the same with a second diode-laser-based DIAL system, this one to take the atmosphere's temperature based on how oxygen absorbs the laser's light. The Lower Tropospheric Observatory would meld the water-vapor-measuring ability of WV-DIAL with the temperature-measuring ability of the $O_2$-DIAL and DLB-HSRL duo (DLB-HSRL, in addition to spotting cloud droplets and other aerosols, is needed to ensure the accuracy of this laser thermometer, Spuler says).

Hayman walks me into the room next to the one in which Eloranta's GV-HSRL resides. He stands on a cylindrical step stool of the sort librarians prefer and points out the heart of the LOTOS system on a rack the size of a baby's crib. Red, yellow, blue, and white fiber-optic cables weave about like fly-fishing line, their mint-green connectors linking small islands of hardware bolted here and there. Boxier hardware fills a rack below it. Spuler, Hayman, and Repasky are building five of these systems at the moment, test articles that will focus just on the water vapor and be stationed tens of miles apart at a to-be-determined location. If all goes well, they'll add the temperature-measuring technology, and the combination's five lasers will blink at the sky and give atmospheric scientists a chance to see if the aerosol, water, and temperature data do in fact improve forecast accuracy.

Carbone is confident that they will, and that a network of something like LOTOS at perhaps hundred-kilometer (sixty-two-mile) intervals would do the trick. It wouldn't come cheap: the price of a new Lamborghini or so per unit, times maybe four hundred across the lower forty-eight. But compare that to the cost of the 150 or so NEXRAD radars in the US network, which run about $8 million each. It wouldn't take many well-predicted hailstorms or twisters to pay off this weather-forecasting investment.[25]

Spuler's and Hayman's creations are among countless lidars devoted to improving our understanding of the air around us and the skies above, all with the common ancestors of Fiocco's and Smullin's borrowed hardware and Hutchie's searchlight dreams. One happens to be a couple of miles away from Spuler and Hayman's NCAR lab. And also in Antarctica.

It's a Saturday afternoon, and Xinzhao Chu is wearing a sweater featuring a penguin. Chu has a thing for penguins, and she's earned it, being among the very few who have seen them outside a zoo. She is a University of Colorado aerospace engineering professor working with the joint CU-NOAA Cooperative Institute for Research in Environmental Sciences (CIRES). She has a gigantic smile and a laugh that comes quick and shakes the room. She has applied her specialized lidar expertise basically everywhere, including Antarctica (eleven times and counting), where two of her graduate students are at this very moment.

Her colleagues' offices are quiet. I have the impression Chu works every Saturday afternoon. She spends much of this one telling me about a lidar quite different from the one Spuler and Hayman are working on. It takes aim where, in Fiocco and Smullin's 1963 follow-up, faint echoes of ruby light returned from what they thought might be meteorite dust.

Chu studies the space-atmosphere interaction region, or SAIR. This starts at about fifty kilometers (thirty-one miles) in altitude, in the mesosphere, the layer over the stratosphere. The Karman line at the top of the mesosphere, about one hundred kilometers (sixty-two miles) up, is where NASA considers outer space to begin. SAIR pays this boundary no mind. It rises through the thermosphere and then into the bottom of the exosphere at about a thousand kilometers altitude (the exosphere finally gives way to interplanetary space at about ten thousand kilometers altitude). SAIR is where the atmosphere slowly gives way to space, and scientists like Chu argue that we know way too little about it.

For many of us, the interactions between Earth's atmosphere and space are not exactly front-burner. We have more pressing things to do, such as thinking about what's for lunch. Of course, there might *be* no lunch without these high-flying layers—had the planet been punished by the extreme solar radiation and cosmic rays they block, we might never have evolved.

Chu's work in lidar started with Chester Gardner, a University of Illinois professor who, like Spuler, came to lidar after a stint in the telecom industry—in Gardner's case, working on early computerized telephone switching systems with Bell Labs in Naperville, Illinois. That was in the early 1970s. Not long before, in 1969, British researchers had taken a newly invented dye laser—which, unlike ruby, could lase different wavelengths depending on the dye—tuned it to a yellow wavelength of 589 nanometers, and lit up atoms in the sodium layer about ninety kilometers (fifty-six miles) up.[26] The phrase "lit up" is key here: this was long-distance spectroscopy, in which the laser light jolted the electrons of sodium atoms to higher energy levels and caused them to glow as they settled back down. The origin of the sodium in that layer was still a matter of speculation, Gardner says: some thought it had burned off from tiny meteors as they cooked in the atmosphere, and others believed it had to do with sea salt rising to incredible heights.[27]

In addition to wanting to solve that mystery, Gardner and others recognized that if you know that sodium resonates at 589 nanometers and the light is shifted away from that, the Doppler effect could tell you about the winds up there, much as Milton Huffaker's lidar could tell you about the winds down here. Precise lidars could also check the temperature, based on the spectral broadening of the sodium absorption lines. Sodium was really a tracer, a Rosetta stone for the mesosphere, where only the occasional rockets could fly.

There was another practical use of lidar lighting up the sodium layer, one that French researchers suggested in 1985. It could be used to enable better ground-based astronomy. Putting big telescopes on mountaintops helped, but the atmosphere still blurred incoming light. The glow of sodium atoms could serve as an artificial guide star, which adaptive optics could use to compensate for turbulence in the atmosphere between the sodium and the telescope.[28] So calibrated, the skies would look as crisp as from a space telescope.

"When I read that paper, I thought, 'By God, I think we can make that work,'" Gardner says.

He and Laird Thompson, a University of Hawaii astronomer, proceeded to light up sodium for telescopes on Mauna Kea.[29] Thirty years hence, you'd be hard-pressed to find a major telescope without a sodium laser guide star to sharpen its vision.

Xinzhao Chu joined Gardner's team as a researcher in 1997. She had crushed her college entry exams in her hometown of Shuangyashan in northeastern China ("It means 'double-duck mountain,'" she explains, writing it out on a sheet of paper, as she did at other times when her accent and my ignorance unhelpfully merged); gone on to study atomic, molecular, and optical physics at Peking University; and done a postdoctoral stint in the German city of Mainz. Gardner was looking for a laser expert, which Chu was, to help build something called an iron Boltzmann lidar (which, if it weren't a lidar, would be a mixed martial arts choke hold).

By that point, researchers had developed fluorescence lidar, as it's called, tuned to excite other lofty metals—potassium, calcium, iron. All of them, including sodium, were clearly from cosmic dust (a.k.a. micrometeors) and not wayward ocean salt. Chu applied her laser spectroscopy expertise to the iron Boltzmann lidar and then from 1999 to 2011 flew with it over the Arctic, Antarctica, and many points between during a grueling but scientifically fruitful "pole-to-pole" lidar expedition that remains a point of pride.[30] Since she moved to Boulder in 2005, her team's innovations have revolutionized the field of resonance-fluorescence lidar and led to many scientific discoveries in the atmospheric and space sciences. Now she has teams of two PhD students spending months at a time at McMurdo Station, Antarctica, where they tend to the upgraded iron Boltzmann lidar Chu first installed at New Zealand's Arrival Heights Observatory in 2010. They stay through the Antarctic winter, taking advantage of the clear, crisp air Hutchie once imagined—and, more importantly, the months-long Antarctic night, during which their lidar operates. They're using these "lidar marathons," as they call them, to understand the nature of the hurricane-speed winds and boiling temperatures from 30–200 kilometers (19–124 miles) up.

Chu describes some of her team's discoveries—of neutral iron in the thermosphere,[31] of mesospheric persistent gravity waves,[32] of the origins of the bulk of our cosmic dust probably being from Jupiter-class asteroids—then picks up the phone and dials Antarctica, which to me seems as far-out as dialing Mars. Not ten minutes later, PhD students Claire Miller and Zhengyu "Harry" Hua are leaning into Chu's laptop screen via Skype. They're both twenty-three. This week, they have had five clear days and, with Miller working day shift and Hua nights, have run the lidar for ninety-six hours straight. I ask how they get the three miles from McMurdo up to the lidar at Arrival Heights when it's minus forty and pitch-dark.

"They walk," Chu says.

"You walk?" I ask the two on the screen.

Chu laughs; the room shakes. No, they take a Ford F-350 that has, as Miller describes it, "really big tires. I'm not very tall, and I, like, need a step-ladder to get into the truck."

Miller has been there for a couple of months and is set to come home. Harry has been there for ten months and won't come home until after he helps train the next batch of winter-over students and helps Chu install a second lidar, this one tuned to sodium, in October.[33] Harry says he misses Chinese food. Miller misses "seeing grass and trees and everything. Being here, it's so hard to remember it's summer in the United States. People will send me pictures of flowers and I'm like, 'What's going on?'"

The word "tough" may not be the first one that comes to mind when you think about scientists. But there's no better way to describe firing lasers through months of Antarctic-winter darkness to discover things like gravity waves two hundred kilometers (124 miles) up. It's tough—intellectually, mentally, physically, psychologically.

**Xinzhao Chu in Antarctica. In the foreground are lidars of her design. (Photo courtesy of Xinzhao Chu.)**

Before I go, Chu talks about the big, big picture. She's interested in connecting the outer reaches of the atmosphere to the lower atmosphere, which is in turn connected to the land and oceans we surface dwellers know best. But there are billions of planets out there, and the processes that shape atmospheres, top to bottom, seem to be universal. The better we understand the most distant reaches of our planet's atmosphere, the more accurately we'll be able to guess what's happening on planets orbiting distant stars based on what we detect from their atmospheres as our space telescopes reach deeper into the galaxy. To that end, Chu, Gardner, and others have proposed a $150 million lidar observatory, one with powerful sodium, iron, and other lidar that can see five times farther than anything ever built.[34]

"We are trying to explore the fundamental processes in Earth-space-atmosphere interactions, which we believe may be universal to other planetary atmospheres that can support life," Chu says. "So I believe this kind of research will help us understand these universal, fundamental processes that could potentially help the exoplanet search."

I thank her for taking so much of her day with me. She gives me names of various graduate students, past and present, with whom she would like me to touch base. She cares about them beyond the narrow bounds of scientific production, it's clear.

Chu's work is but one example of the countless ways creative people have applied lidar in the study of the atmosphere we owe our lives to. And lidar is used in a lot more than atmospheric science. I walk back outside and make a point of appreciating the warmth, the trees, and the flowers.

# CHAPTER 5
# TAKE THE PLUNGE

Lidar uses a laser as a tool to measure things. A new measuring stick first gets used in places where old measuring sticks don't reach, don't do a good enough job, or are cumbersome or expensive. In atmospheric science, lidar pioneers aimed their lasers at extreme altitudes, amorphous toxic plumes, and invisible winds that threatened airplanes and the people in them.

Bathymetry, a field dedicated to measuring land that's underwater, presented another opportunity for a new measuring stick. Perhaps the laser could finally illuminate the mysteries of the depths—or, more accurately, the shallows. Doing so could have enormous practical implications. Lidar's pioneers, from Hutchie to Herb Ligda to Xinzhao Chu, drifted into lidar from other fields, drawn in by this new measuring stick's potential. So it was with George D. "Dan" Hickman. Hickman, a high school track star from Ypsilanti, Michigan, graduated from the University of Michigan with a PhD in nuclear physics in 1959 and worked in nuclear power for General Electric and the Swiss government. In the mid-1960s, he moved to Syracuse University Research Corporation and drifted into radar studies. This was (and is, under the name SRC Inc.) a contract research house like SRI.

"We were around a lot of water up there in Syracuse," Hickman recalls. "I had some conversations with some of the professors up there, and I decided, 'Why not try the laser in water?'"[1]

Starting in 1967, he landed a succession of small contracts from the US Navy to look into how laser light behaves underwater. The idea at first was to look into underwater communication. But that mandate soon changed. The navy needed a better way to map not only shorelines and beaches but also the submerged land near the shore. Hickman's Office of Naval Research sponsors put it in clear terms. Amphibious landings demand a detailed knowledge of length of usable beach, beach width, beach gradient, beach approach, surf and tidal range, beach material, nearshore current, offshore bars, and more. Wind and water can change these places quickly (compared to, say, a mountain range), making beaches and the earth hidden under waters approaching them "the most dynamically complex of all land forms."[2]

The world's great sea power needed to understand the nature of coastlines all over the world, and sooner rather than later. Without better bathymetry, the navy's mobility and flexibility were "critically impaired," and the element of surprise, a staple of amphibious landings since before the Trojan Wars, was at risk.[3]

The potential of bathymetric lidar had not been lost on civilians, either. An aircraft with a lidar would survey a lot faster than a sonar vessel would, and it could cover shallows or risky waters that boats couldn't reach. Lidar scans could help planners align underwater pipelines and could support oil and gas exploration, help predict storm surges, and assess storm damage. Lidar could help set the baseline for jetty and breakwater projects and help keep ports and harbors safe. It could help manage fisheries and coral reefs. In short, lidar seemed to have a bright future in the water—if it could be made to work.

The old-school bathymetric tool was a rope with its lengths marked off and a weight to pull it until it hit bottom. By the 1960s, it had given way to shipboard sonar, which used sound waves to determine depths. But the ships carrying the sonar couldn't go all the way to the shoreline, much less take in the beach. In early 1966, the Vietnam War was heating up, and the navy needed to survey hundreds of miles of South Vietnamese rivers and shallow coastal waters. They relied on slow-moving sonar boats within range of enemy territory. Aircraft-mounted lidar presented a tantalizing alternative. Aircraft could cover a lot more area a lot faster, and it might work both above and below the waterline.[4]

Others had experimented with lidar in water. Out at SRI, for example, the navy was paying for a project to see if lidar could spot submarines.[5] But Hickman would be the first to fly a lidar aimed into the water. He started on the ground, borrowing from Avco Everett Research Laboratory a laser that shot blue-green blips. He shot it in a twenty-foot-long tank of Lake Ontario water to see how the surface reflected the laser light and how the water itself absorbed and scattered it, among other things. Blue-green lasers penetrated water, but no one knew how well.

As a nuclear physicist dabbling in lasers, Hickman wasn't clear on how to mount a laser to a Turbo Commander aircraft when the time came in 1968. He enlisted Syracuse Research colleagues with experience in airborne radar to help shock-mount, bolt, and otherwise tie down the bulky laser and various electronics, pumps, and power supplies interconnected with cables as thick as garden hoses. They jury-rigged a mirror at forty-five degrees in front of the laser, which sent the outbound pulses through a hole in the bottom of the fuselage. The receiving telescope got its own hole. They flew about fifty miles of Lake Ontario coastline, from Sodus Point to Sandy Pond, over a few nights. They stuck to an altitude of around five hundred feet, staying close to shore, recording depths of up to twenty-six feet on an oscilloscope with a screen the size of a Post-it Note. The jagged traces were recorded with high-speed Polaroid film. Comparing his results to navigation charts, Hickman estimated the system to be accurate to about a foot and a half vertically. A lidar flying faster with a better laser could, at three hundred miles in an hour, cover as much as fifty square miles an hour, Hickman calculated.[6]

The next year, Raytheon delivered to the navy a prototype bathymetric lidar called the Pulsed Light Airborne Depth Sounder, or PLADS. Navy researchers mounted it on a helicopter and tested it on the Patuxent River and in the waters off Panama City, Florida, among other places, managing depths of more than 90 feet in Florida's clear waters. In the muddy Potomac, the return signal went dark at about 18 feet.[7] Around the same time, the Naval Air Development Center's enormous helicopter-mounted lidar managed to get blue-green photons back from 225-foot depths near Key West.[8]

Two hundred and twenty-five feet of visibility may be a smoggy day in Beijing, but it's great in water. The challenges of using bathymetric lidar extend well beyond turbidity. If you're aiming at a smokestack plume a mile away, determining its shape to within a few meters is fine. A few meters in bathymetry is the difference between smooth sailing and a shipwreck. So your lasers and detectors have to be more accurate, which means they have to be faster. The nature of water and the particulates suspended in it complicate things too. Calm seas can act like a mirror; waves scatter light and broaden laser pulses; and water and the stuff floating in it block and deflect photons, so detectors have to get by with fewer returns. Then you've got the added complexity of getting it all to work in an aircraft.[9]

Dan Hickman moved on to a company called Sparcom but kept working on lasers in water. The navy, NOAA (responsible for coastal navigation and charting), and the US Geological Survey (responsible for mapping the country) paid for much of the work. In a big tank, he added small increments of various marine sediments collected from shallow-water sites along the East Coast, systematically increasing the water's murkiness to estimate how well blue-green laser pulses of slightly different colors and energy levels could penetrate it. He also did theoretical work to establish the best scan angles, patterns, colors, and laser spot sizes from an aircraft.[10] He'd soon have his hands on an airborne system again.

By late 1973, PLADS was gathering dust and ended up at what's known today as NASA Wallops Flight Facility on Virginia's eastern shore. There, NASA scientist Hongsuk Kim considered using it as a basis for a new system he called the NASA Airborne Bathymetry System. PLADS had been tetchy at best; with help from NASA's Langley Research Center, which had done serious work on atmospheric lidar, Kim decided to start fresh. The new system would involve different lidars doing different things: not only would it do bathymetry, but it would also zap oil slicks and algae with ultraviolet light powerful enough to cause the targets to glow like Xinzhao Chu's iron and sodium atoms in the mesosphere.[11]

Kim brought in Hickman to work on the fluorosensing side. Among the tests was a June 1973 flight in NASA's DC-4 aircraft that buzzed four hundred feet above the Delaware River. This was a prop job that carried eighty passengers as a commercial airliner, and Kim and Hickman needed the space because the hardware would have been a tight squeeze in anything much smaller. The US Coast Guard dumped four hundred gallons of no. 4–grade heating oil into the river for the lidar to detect, which it did, to the point that it could see "extremely low-level oil in water which cannot be identified by ordinary photographic method."[12] Over the next year and a half, Kim and colleagues flew their lidar over Key West's Boca Chica Bay, capturing their results on a camera that photographed oscilloscope traces, as well as on a digital system that recorded on reels of magnetic tape. This was a leap forward in what would become lidar's endless game of data-processing catch-up. The system also took a trip to Lake Ontario, where the team found that the United States' side of the lake was more polluted than the Canadian shore.[13] The lidar worked just well enough that the navy, the Defense Mapping Agency, NOAA, and NASA teamed up to back a second-generation version called the Airborne Oceanographic Lidar, or AOL.[14] A NASA Wallops team led by Frank Hoge did the heavy lifting. While it may be an exaggeration to say that AOL would be to airborne lidar what the Wright Flyer was to being airborne, it's not much of one.[15]

Gary Guenther, a NOAA scientist, played an important role in processing and interpreting the first AOL bathymetric data, though he never actually flew with the system he helped make work. Born and raised in a small town in South Dakota, he had won a scholarship to Northwestern University, where he earned engineering and physics degrees in his undergraduate and master's programs. By the time he got to NOAA, he was in his early thirties and happened to have worked with green lasers for the military. His boss, who had gotten wind of NASA's proposal to build AOL, asked him what happens when a laser beam goes through water.

"I thought about it for a second, and I said, 'Give me a couple of hours,'" Guenther says. "I was still working on it thirty years later."[16]

He did his AOL work via a terminal on his desk in Riverdale, Maryland. It connected to a mainframe in Mississippi through dial-up modem. The crux of the challenge was to make sense of the peculiar interplay of man and nature that marks nearly all lidar. Man creates pulses of light by driving electricity through clever combinations of metal, plastic, and glass (and, with some lasers, gases). The pulses are imperfect to start with; nature musses them further. In the case of bathymetric lidar, it does so with varying light conditions, winds, waves, swells, depths, water clarity, bottom composition, and so on. Man captures some fraction of the original pulses and converts the dancing light waves into digitized representations called waveforms. In the case of AOL, man then gave the digitized-waveform outputs—countless strings of numbers representing hour-long flights that captured four hundred laser-pulse waveforms a second—to Guenther, whose job was to use them to figure out where the surface was and where the bottom was.

"That's when the rubber meets the road—when you get the data back from the aircraft. The data is contaminated with problems of the digitizer, with the environment, with distortions and mistimings that come from you don't know where," Guenther says. "Then it's like, 'Oh, my god, how do I get the right answer out of this total piece of shit,' if you'll pardon the expression. And that's why it took thirty years."

This was not to malign the creators of AOL, who built an incredibly complex system from scratch in an era when a mainframe computer had a tiny fraction of the processing power in a modern smart watch. And it ultimately worked. Guenther and others, through iterative loops of field testing, determining what was what, tweaking algorithms, and adjusting hardware, figured out how to interpret the complex signals from different water clarity, depths, and sea-bottom characteristics into reasonably accurate readings on where the surface was and where the bottom was. They developed automated depth algorithms and statistical routines and encoded them into computers that filled rooms. Feeding those computers were computers the size of refrigerators that flew along in the NASA C-54 aircraft. The airborne system, including a two-thousand-pound laser table, lasers, and electronics, weighed close to three tons and filled most of the four-engine plane. This was more invention than engineering—applied physics, really. They wanted to know whether such a system could work for producing nautical charts and doing coastal surveys, and whether it could ever hope to work at a reasonable cost without PhD-level technical talent along for the ride.[17]

In 1977, AOL flew eighteen missions on either side of the Delmarva Peninsula on Maryland's Eastern Shore: over rivers, bays, and ocean waters; in summer and winter; under clear and cloudy skies; during the day and at night; in windy and calm weather; through clear and murky water. It amassed five million soundings, and it led to advances in mapping lidar that extend far beyond bathymetry.[18]

Meanwhile, other governments were sponsoring other bathymetric lidars.[19] One with far-reaching impacts happened to be in Canada.

The house in the old neighborhood twelve miles north of downtown Toronto is new. Allan Carswell answers the door. He is tall, lean, white-haired, and friendlier than a man of his accomplishments needs to be. At 84, he's spent a lifetime applying his formidable intelligence to extraordinary problems. Prior to my visit, I asked about interviewing his wife, Helen, who had, with him, built the company I was here to talk about. A black-and-white photo of her leans on a shelf in his office just inside the door. She's young and dark-haired and beautiful in her nurse's uniform.

Carswell walks me to a den down a short hall. Soft music plays. Helen sleeps in a recliner under a blue fleece blanket with a floral pattern. This is her life now, he tells me. His wife of sixty-one years hasn't spoken for the last five. Before then, they had razed the house they had lived in for forty years and rebuilt on the same spot, moving their bedroom to the first floor and designing with an eye toward making it easier to care for Helen as her Alzheimer's advanced.

"We designed what we called our 'geriatric house,'" he says.

We sit in a bright living room in the geriatric house and talk about history that far predates his pioneering lidar work at York University and his and Helen's founding of Optech in 1974. We start with his father, Duncan, the son of a Scottish butcher, who enraged his family when he ditched Bannockburn for the Royal Navy after the tenth grade. His classmates ended up conscripted into the World War I trenches, and as he told his son many years later, "they were dead within six months." He came to the United States, working as a butcher in Chicago and then Toronto, where he met Carswell's mother. Margaret MacAskill had come from Scotland's Isle of Lewis, arriving in Toronto on a government program in which she would repay the £15 she was given up front through her work as a maid. They had three kids; the second, Allan, would be the first in the family to go to college.

He studied engineering physics at the University of Toronto and stuck around for his PhD in high-pressure physics, which has to do with gases so squeezed that they act like liquids. He spent a year at the Institute of Theoretical Physics in the Netherlands, then landed in Montreal at RCA Victor Research Laboratories in May 1961, where he joined the Plasma Physics Laboratory. There, he studied how the plasma—superheated gas—coming off intercontinental ballistic missiles would look on radar and, by extension, how to determine the shape of the underlying object. With the laser having recently been invented, Carswell's managers decided to have the new guy build one. So Carswell did—Canada's first helium-neon laser, which is essentially a plasma that shoots out a laser beam. A carbon dioxide laser followed, one that, as he puts it, "could burn holes in fire bricks." Soon he was leading Canada's first corporate laser R&D group, until he left it because of a broken leg.

The leg wasn't his. The Carswells' two-year-old daughter "was horsing on the escalator in a big store, and Grandma went to grab her and stumbled," he says. Helen insisted on moving back home to keep an eye on her mom. That happened in 1968, the year Carswell joined York University. The university, only a decade old then, was launching a science faculty. He came across the work of Fiocco and Smullin and others who had bounced laser light off the moon, pollution plumes, and more. He had spent years bouncing microwaves off plasmas and other things.

"So I just thought, well, maybe I'll try lidar," Carswell says.

He borrowed a ruby laser from one of his new colleagues and fashioned it into a lidar system, then got grants for a better lidar and a box truck to transport it and its hefty electronics. It was the second mobile lidar he was aware of, the other being that of the SRI team in California. Like SRI's, Carswell's lidar looked into the air, and he worked with Ontario Hydro and others to track plumes from the 650-foot-tall stacks of the utility's new, enormous coal-fired Nanticoke power plant. The lidar saved the utility from having to fly around blindly in helicopters sniffing for pollutants.

Ontario Hydro liked the system so much the company wanted one of its own. Allan and Helen spent the 1973 Christmas holidays weighing the pros and cons of starting a company to build such a thing. "What really drove me was I recognized that there was one hell of a great potential in this technology, and at that time I was one of the few people that knew about it," he says.

They decided to do it, but with Carswell staying on at York. They incorporated in 1974; Helen left nursing and took a typing class to help with proposals. Their first hire was Sebastian Sizgoric, who had recently worked for Carswell at York. Optech's bid was lower than that of the only other proposal for the Ontario Hydro system—from SRI—and Carswell promised to deliver sooner. Ontario Hydro was concerned enough that the upstart might not stay in business that the utility insisted on buying all the hardware and doing the fabrication work in its own shop.[20]

This was fine with the Carswells. For the next several years, Optech was a contract research house, bootstrapping one project after the next, with Carswell and Sizgoric chasing ideas from various corners. They looked for natural gas pipeline leaks; they created an X-ray monitor for use in hospitals; they developed an underwater lighting system for turbid waters; they proposed laser-generated airplane cockpit displays, lasers for the construction business, lasers for cutting begonias at a commercial nursery, lasers for seeing bubbles in glass for a glass maker, lasers for seeing through battlefield smokescreens. They had Randy Bachman of the band Bachman-Turner Overdrive over to check out a prototype system for laser light shows, called LaserVibe.

**Allan Carswell working on a lidar he built for the Canada Centre for Inland Waters starting in 1969. (Photo courtesy of Teledyne Optech.)**

One of the big things that stuck was bathymetry. Carswell had been pointing lasers into the water since 1969, when he and his York team built an argon laser to understand how murky water in tanks at the Canada Centre for Inland Waters scattered the light. By 1973, they had a follow-up system and tested it on Centre research vessel Limnos out on Lake Erie. The Canadian Hydrographic Service and the Canada Centre for Remote Sensing saw enough potential to enlist Carswell and Optech to push ahead on a possible airborne version of the system. In 1978, Optech flew its first lidar bathymeter, upgrading it as various enabling technologies improved over the course of several years. By 1984, Optech had built the first operational lidar bathymeter, the LARSEN 500 (AOL was technically a research system). It scanned back and forth as it fired twenty laser pulses per second. The Canadian Hydrographic Service had sufficient trust in the technology to produce the world's first airborne-lidar-derived hydrographic chart, #7750. It sounded out Cambridge Bay in the Canadian Arctic, which was shallow and hard to get a sonar boat into, and where ice only relented briefly each year, making for brief operational windows. A note on the map reads: "Data is from a Light Detection and Ranging (LIDAR) survey. Soundings from various older and less reliable sources have not been used."[21]

What followed was a long march of better performance in generally smaller packages, with applications for undersea mapping as well as military uses. A system delivered to the Swedish Defence Institute was to "detect underwater objects." One built for the US Defense Advanced Research Projects Agency (DARPA) was to detect underwater mines. It involved a fifteen-foot-long copper vapor laser that operated at 1,500 degrees Celsius, hot enough to melt steel. Optech delivered it on a hundred-day crash program during the first Gulf War.[22]

Gary Guenther, meanwhile, stayed at the algorithmic intersection of man and nature. When AOL wound down in the early 1980s, NOAA let Guenther work on a navy system called the Hydrographic Airborne Laser Sounder, or HALS. This was a fancier system than AOL, with separate green and infrared laser beams.[23] Guenther's couple-of-hours project had now stretched to a decade, with no end in sight. Along the way, he wrote an inch-thick manual about what he had learned that was casually referred to, depending on whom you asked, as either the "Blue Book" or the "Bible."[24] Driving it all was Guenther's belief that anybody could shoot a laser into the water and get a return. That wasn't the point. "Your job is not surveying. It's surveying accurately," he says.

The quest had Guenther working all hours. The annual family vacation to Capacon Resort State Park in West Virginia saw many days with his wife and kids out at the lake as he sketched waveforms and wrote Fortran code on yellow legal pads back in the log cabin. The navy system wasn't cooperating. Despite millions of dollars invested and good initial results, each flight seemed to bring back sketchier data, which meant Guenther had to invent new ways to poke and prod the results into surface and bottom returns that met the standard of "surveying accurately." The system, it turned out, was being shaken to death in the aircraft carrying it. The navy higher-ups lost patience and killed the program.[25]

Guenther had codified his lessons from AOL in the Blue Book, and so his insights had propagated to efforts in Canada, Sweden, Australia, and the Soviet Union. But he had made real progress on this now-deceased navy project—algorithms capable of surveying accurately despite all sorts of common hardware foibles and natural vagaries. His progress might have evaporated with HALS had it not been for a phone call out of the blue from a "young kid" down in Mississippi.

It was 1986, and Guenther was in his forties now, old enough to refer that way to a US Army Corps of Engineers staff researcher in his late twenties. His name was Jeff Lillycrop, and he worked out of Vicksburg, Mississippi. His managers had asked him to look into lidar. Among the corps's many responsibilities was and is to make sure America's roughly thirteen thousand miles of shipping channels stayed wide and deep enough for ships to pass through. We may like to think of ourselves as a digital society, but more than half of US imports and about 40 percent of our exports come and go on ships. We're really a maritime society.[26]

"About everything you buy at Walmart comes to the United States on a boat," Lillycrop says. "Bathymetric lidar was identified as a possible technology that could help us survey our navigation channels better, faster, cheaper."[27]

Shipping channels are constantly filling with sediment and have to be dredged. The corps was using a fleet of sonar-toting boats to monitor them. At that time, the sonar pinged straight down, returning a single point from the bottom. The boats ran up and down a channel to piece together a rough idea of the state of sedimentary affairs. Lillycrop had heard about Optech's LARSEN-500. He also knew that other government agencies had reasons to be interested in lidar—in particular the US Navy, as well as NOAA, which was responsible for US nautical charting. He knew about Gary Guenther and his Blue Book/Bible and figured that if anyone had an idea about how to proceed with airborne bathymetric lidar, it would be him.

"I didn't laugh him off the phone, but I'd been trying for like ten years to sell an airborne lidar system to NOAA," Guenther says. "Unsuccessfully, because they wouldn't put in the money and were still led by a bunch of ship drivers."

NOAA was only part of the challenge, Guenther well knew. Lillycrop would also have to get the US Army Corps of Engineers and its various sonar-steeped coastal districts on board, not to mention the navy, which was bruised from the HALS fiasco. "None of this was possible," Guenther says. "And none of this would have ever happened had it not been for this young punk kid who didn't know that he couldn't get millions of dollars to build a lidar system and get it out of the R&D lab and get it into the corps districts. It was crazy."

What drove Lillycrop was that the civilian and military potential of airborne bathymetric lidar was too great collectively to ignore individually. He set up a meeting with Guenther in Washington, DC, which led to conversations with Optech's Carswell and Sizgoric as well as the Canadian government, which led to detailed study on the viability of building a bathymetric lidar capable of meeting everyone's needs. A year and a half later, a three-inch-thick report detailed how such a system could work. The partnership became the Joint Airborne Lidar Bathymetry Technical Center of Expertise, or JALBTCX. Lillycrop had envisioned it; now he was leading it.

Optech delivered the three-inch-thick report in 1989. That it took four more years to finish the system gives some indication of how hard it was to do. Guenther's work on the ill-fated HALS system found a home in the new product, called SHOALS. At first, it wasn't fast enough to fly on anything faster than a helicopter.[28] But it satisfied the International Hydrographic Organization's requirements for use in charting, and successive versions sped up and got increasingly accurate. With time, SHOALS also extended its reach onto the beach, turning the bathymetric system into a topobathy system (*topo-* for topographic, or land-elevation mapping).

The irony is that the early idea behind the creation of SHOALS—keeping tabs on shipping channels—never quite panned out as Lillycrop had thought it might. The murk in shipping channels, particularly in back bays, shrouded the depths, for one thing. And the system's speed led to an unexpected outcome: capturing the length of a channel in clear waters could be done so fast that the couple of days' setup time and the higher cost of running the airborne system made it uneconomical, particularly as vessel-based sonar systems improved enough to take the full swath of a shipping channel in a single pass.

But then another door opened. SHOALS was, Lillycrop says, "about the only system in the world capable of near-shore coastal mapping." So rather than overflying shipping channels, SHOALS was sent around the US coast, scanning five hundred meters onto land and one thousand meters out into the water—about a mile-wide swath. Today, the program maps the entire continental US coast under the auspices of the NOAA National Geodetic Survey's Coastal Mapping Program, repeating it every five years. Also, they re-fly places that have been hit by major storms to assist with assessing, for example, Hurricane Irma's impact on the Florida coast in 2017, which feeds into stabilization and recovery efforts.

Mapping the coasts has helped manage shipping channels, Lillycrop says, because it provides a big-picture sense of how coastal sediment moves and, more broadly, how a range of human and environmental processes—storms, floods, currents, condo towers—affect the underwater landscape. "The more we understand those processes, the better off we are," he says. JALBTCX now flies its lidar over wetlands and flood-control projects too, he adds.

A five-year, $13 million program that wrapped up in 2011 upgraded and reimagined SHOALS as a system called CZMIL. It can see deeper and penetrate through cloudier water, shoots ten thousand pulses per second, and includes a high-resolution camera and a hyperspectral imager.[29] The US Army Corps of Engineers and the navy now share four such units. The navy flies CZMIL over distant shores, where its surveys set the stage for better military and humanitarian planning and, more generally, boost allies' hydrographic surveys to international standards, which helps ensure the safe passage of not only naval vessels but also container ships and tankers carrying many imports and exports.[30]

This is all a long way from Dan Hickman's maiden flight over Lake Ontario back in 1968, or Allan Carswell's trolling about Lake Erie with his big laser on the Limnos. Carswell retired a few years ago to focus his attention on caring for Helen and leading the Carswell Family Foundation, which gives away the money his lidar expertise and business savvy earned him. The company he and Helen had bootstrapped since 1974 had grown to three hundred employees and diversified well beyond bathymetry by the time Teledyne bought it outright in 2015.

"I would say that everything that Optech is doing nowadays wasn't even a gleam in your eye," Carswell says. "When we were out on Lake Erie, we could get down maybe twenty meters and be able to measure depths and turbidities and stuff like that, but to be able to do that at aircraft speeds by the square kilometer, there's no way you would say that was possible."

# CHAPTER 6

# DISCO BALL IN SPACE

James Faller graduated from the same high school in Mishawaka, Indiana, as his parents had. They had been at the top of their class, as was he. But Faller had the good fortune of not coming of age during the Great Depression, so rather than going on to manage an ice cream store like his dad did, Faller went to Indiana University down in Bloomington, where he studied physics. As he wrapped up, a professor suggested graduate school, and then suggested Princeton.

"Where is that?" Faller asked.

"That's where Einstein is," the professor said.

Faller started at Princeton in June 1955. He got there too late for Einstein, who had died two months earlier. "I did learn that he drank Earl Grey tea, so I tried it for quite a few months," he says. "But it didn't help."[1]

The link between Earl Grey tea and Einsteinian genius remains tenuous. But Faller, like Einstein, was interested in gravity, and his PhD work involved designing modern twists on the apocryphal Galileo experiment of dropping balls of different weights from the Leaning Tower of Pisa. A couple of years in, he asked his advisor, Robert H. Dicke, about having regular after-dinner meetings of a half dozen or so grad students and postdocs on Thursday or Friday nights. It became a cross between an academic salon and an open confessional booth, a mix of troubleshooting and fresh ideas. One evening in 1962, Faller brought in a fresh idea. It was in a twenty-five-cent red-and-white rubber ball he had bought at Woolworth.

Faller had cut the ball open and glued a corner cube retroreflector inside. A corner cube is what it sounds like: three mirrors arranged to create a corner. Light coming in reflects right back to the source, even from odd angles. "The Mona Lisa of the world of optics," Faller calls it. "No matter where you are in the room, it looks right back at you." He soldered two thick wires together and made what he later referred to as a "nose guard" over the hole. The ball, bottom-heavy from the retroreflector inside, could roll out of an unmanned spacecraft on the moon and come to rest with the opening facing up. Light beamed up from Earth would reflect right back to the source.[2]

That could establish the Earth-moon distance much more accurately than had been possible with radar, which had narrowed it down to within about a kilometer. More interesting to Faller was putting Einstein's general relativity theory to the test. If you could measure with precision the Earth-moon distance and how it subtly changes over time, you could treat the planet and its dance partner like enormous versions of differently weighted spheres dropped from a leaning tower. The Earth and the moon are always falling toward the sun, after all. Central to general relativity, the gorilla of gravitational theories, is the idea that mass distorts the fabric of space-time like a ball sinking into a rubber sheet, affecting the motion of everything near it as well as how time flows by. Different theories predict different nuances in the distances between Earth and the moon. The challenge was measuring gravity (or time) precisely enough to prove Einstein right or wrong.

Faller typed up his rolling-retroreflector idea in March 1962, two months before Fiocco and Smullin did their Luna See experiment. Dicke was interested—he had, a couple of years earlier, coauthored a paper discussing ways to track satellites, the study of gravity being the main attraction.[3] That had been before the invention of the laser, but one of the ideas was to make a ten-centimeter-diameter (four-inch) corner cube satellite, fly it at an altitude of about 1,250 miles, hit it with a pulsing searchlight, and take photos of it against the stars. The corner cubes would return five thousand times more light than a metallic sphere the same size.[4]

The timing was good. The Soviets had launched Sputnik in 1957, and the United States had created NASA in 1958. Dicke had done some work with an MIT graduate student, Henry Plotkin, who had been drafted in 1954 into the US Army Signal Research and Development Labs in New Jersey, not far from Princeton. Plotkin moved to NASA's Goddard Space Flight Center in 1960 to develop systems for photographing satellites and stars to calibrate satellite-tracking radar. Before he left for Goddard, Plotkin met with Dicke, who urged him to pursue his idea of using corner cubes to track satellites. Plotkin did, looking into satellite designs, corner cube sizes and array configurations, searchlight specs, and how much light would be needed for the corner cube reflections to show up on film. Dicke lobbied NASA, and support grew quickly. The big break came when the manager of Goddard's new Beacon Explorer program agreed to add corner cubes to satellites being designed to send radio signals of varying frequencies to map the ionosphere. By the time Faller brought his Woolworth rubber ball to that meeting in Princeton, Plotkin had an approved proposal to zap Explorers, one that substituted a laser for the searchlight.[5]

Plotkin had a team ready when the first of the three Beacons launched in March 1964, but a rocket problem sank that spacecraft in the Atlantic off the Brazilian coast. Beacon Explorer B, also called Beacon 22, launched that October and settled into an orbit about a thousand miles up. By then, Plotkin and a small team had set up at the Goddard Optical Research Facility out on a hill in the fields northeast of Goddard's Greenbelt, Maryland, nexus. A couple of weeks later, Plotkin and a colleague teamed up to joystick a ruby laser mounted to a Nike-Ajax missile radar tracking mount. With one controlling altitude, the other azimuth, they captured the first-ever satellite laser ranging returns in the darkness of Halloween morning, 1964.[6]

"We didn't blind any witches in the process," says the most important figure in the history of the field that grew out of that experiment. John Degnan was working on Plotkin's team at the time as an eighteen-year-old co-op student from Philadelphia's Drexel University. His role then wasn't terribly glamorous. "I have this image of the young John Degnan lying on his back under a Minitrack camera, installing the 8 × 10 film cassette. (That's what co-op students do.)," Plotkin told an audience decades later.[7]

That was fine with Degnan. He hired into NASA Goddard full-time in 1968 and retired from NASA in 2003 to become chief scientist at Sigma Space Corporation in Lanham, Maryland, a few miles down the road from Goddard. He meets me in his office there. The combination of walls adorned with awards and a desk piled with papers offers solace to all whose desks are piled with papers. There is a can of cream soda. He wears glasses and an orange button-down and gives the general impression of a guy you might run into in a local bar and strike up a conversation about the Phillies. We strike up a conversation about satellite laser ranging.

Scientists around the world quickly recognized the many things SLR, as it's called, could do. You could track satellite orbits far more precisely. You could study gravity. And you could improve the big-picture understanding of our home planet by using the knowledge of where a satellite is to pinpoint the location of the ground station shooting laser light off it. Taken as a whole, SLR was a good way to help establish baselines—reference frames, they're called—against which to measure other things as related to planet Earth writ large. The shape and rotation of the Earth, neither of which are constants if you look close enough, are two examples. We have had twenty-seven leap seconds since 1972, for instance, because one Earth rotation takes very slightly less than the 86,400 seconds we say it does. Sea level rise is another example.

Sea levels are rising at an average rate of about three millimeters (an eighth of an inch) a year. SLR doesn't measure sea level rise per se, but it has enabled the measurements in two ways. One was to precisely establish the orbits of sea-level-measuring satellites. The satellites wear retroreflectors like sequins for this express purpose. The other way was to establish the geoid, which imagines calm, tideless seas across the planet, slicing through the continents, smooth as a billiard ball (though undulating gently with gravity differences and bulging in the middle as our planet does). Gravity is the central player in calculating the geoid.[8]

Satellite laser ranging took off as a scientific pursuit. By 1967, six satellites had corner cubes, and lasers were pinging them from sites in France, Algeria, Greece, and New Mexico as well as the Goddard site in Maryland. NASA and others around the world built both fixed and mobile systems. Part of the draw was plate tectonics: SLR systems provided a great way to measure the motion of continents. NASA placed SLR stations on either side of the San Andreas Fault and sent units to southern Europe for similar reasons. Stations sprang up in Peru, Brazil, South Africa, Poland, Latvia, Bolivia, Cuba, India, Egypt, and elsewhere. The Soviet Union launched its own corner-cube-carrying satellites and built more than two dozen SLR stations to track them.[9]

By 1975, France had launched the first satellite whose sole purpose was to reflect shots from SLR stations—a 9.4-inch ball called Starlette, glittering with sixty retroreflectors. A NASA team led by David E. Smith outdid this the following year, launching an aluminum and brass disco ball of a satellite nearly three times bigger, just in time for the Bee Gees. The two-foot-diameter LAGEOS had 426 corner cubes. More important than its size were its heft (nine hundred pounds) and its orbit: a near-perfect circle at about 3,700 miles—more than six times higher than Starlette's. Up there, LAGEOS would encounter little atmospheric drag. It was designed to be nothing more than, and nothing less than, a glittering, rock-solid reference point in the heavens, one swayed only by forces it was designed to appraise. It also had something cool hidden inside: a plaque Carl Sagan designed to be comprehensible to whatever intelligent life happens to be roaming Earth when the gravity that LAGEOS helps measure pulls it back to the planet in a few million years.[10]

Degnan, meanwhile, went on to work on things somewhat more complicated than switching out film canisters. He earned his master's degree and PhD at the University of Maryland while working at Goddard, becoming an expert in both lasers and detectors. He did much to improve SLR accuracy—from the three meters of the 1964-era system to the ten-centimeter range by the late 1970s to about one centimeter today. Along the way, he worked on a famous experiment with his advisor, University of Maryland physicist Carroll Alley. It involved six atomic clocks in a trailer at the US Navy's Patuxent Naval Air Test Center and another six on a navy submarine-hunting aircraft. The plane flew for fifteen hours at a time at about thirty thousand feet, in a racetrack pattern over Chesapeake Bay. The idea was to test Einstein's relativity, which said the clocks in the plane should run very slightly faster than the ones on the ground due to the slightly weaker gravity field at altitude. Degnan and technician Steve Davis were charged with modifying a fickle laser the navy had donated to the cause.

Degnan and Davis got it to produce the ultrashort pulses (just one hundred picoseconds, or trillionths of a second, long) the experiment demanded. From a tracking mount on a trailer, Davis manually aimed and fired bursts of thirty shots per second at the aircraft about every three minutes. On the plane was a corner cube mounted right next to a window. A detector inside the window recorded the laser pulse's arrival at the aircraft; the corner cube reflected it back to a detector for the atomic clocks at Patuxent (the pulse was wide enough to hit both). The team then compared the time on the atomic clocks in the air with that of the ones on the ground. Sure enough, the ones in the air ran forty-seven billionths of a second ahead of those on the ground by the time the plane landed again, just as Einstein's equations predicted.[11] The laser became the subject of Degnan's 1979 PhD dissertation. If all this seems esoteric, consider that without corrections for relativistic effects, your phone or smart watch's GPS accuracy would slip by about eight meters (twenty-six feet) *per minute.*[12]

Which brings us back to Woolworth rubber balls and corner cubes on the moon. Because it happens that Degnan's advisor, Carroll Alley, had been a gravity-studying, atomic-clock-making Princeton PhD student of Robert Dicke's just a couple of years ahead of James Faller; and that Alley had been in some of those very same evening physicist salons; and that as the American space program sprinted toward humanity's date with the moon before the end of the 1960s, Alley was the principal investigator of what became known as the Apollo Lunar Laser Ranging Experiment, or LURE.

From the outset, Dicke, Alley, and others worked to convince the Apollo program's leaders that having astronauts deposit corner cubes on the moon was good, cheap science. Faller, by then a postdoctoral researcher at JILA at the University of Colorado, came up with a design, the key challenge having been to decide the right size of corner cube. He realized that a corner cube that was bigger than about twelve centimeters (almost five inches) across would require separate telescopes for sending and receiving laser pulses. A big retroreflector would mean a narrower return footprint. The motions of Earth and moon during the intervening 2.5 seconds would land the return signal a mile away. So you'd want smaller, less focused retroflectors so a collocated receiving telescope could sniff out photons on the flanks of the return pulse. Faller considered other factors too, such as how solar radiation or the stark temperature changes that come with the lunar day and night would distort cubes of different makeups. He settled upon a slew of 1.5-inch cubes.[13]

Faller took a faculty job at Wesleyan University as the work continued. For a time, prospects dimmed as a bundle of scientific experiments the astronauts would do on the moon fattened and shoved LURE to the sidelines. Then in September 1968, Faller and colleagues got word that the bundle had grown too big for the astronauts to handle without spending too much time outside their capsule. Placing a retroreflector on the moon would be quick work, and now there was space for it. On July 20, 1969, Buzz Aldrin and Neil Armstrong set the array of one hundred corner cubes down on the moon's fine sands and went about the rest of their giant leap for mankind. Now it was up to lasers and receivers on the ground to find an eighteen-inch-square array of fused silica from 240,000 miles away.

**The Apollo 11 retroreflector array, a small target from 240,000 miles away. (Photo courtesy of NASA.)**

Alley had a team ready at McDonald Observatory in Texas, which would be the initial center for the long-term lunar laser ranging effort; Faller's group of Wesleyan undergrads was at Lick Observatory in California as an insurance policy in case Alley's team came up empty. Henry Plotkin of Goddard, who was on the LURE science team along with Dicke and others, had loaned Faller a satellite-zapping laser as a backup to one Faller had had custom-built for the occasion. For days, the teams in Texas and California searched to no avail. There were confounding factors: the moon was so low in the sky that they had to take a few days off, and humidity over McDonald Observatory soaked up their laser's light. The numbers weren't on their side to start with. They sent up some 10,000,000,000,000,000,000 photons per pulse; they hoped to get maybe a handful of them back.

Major telescopes such as Lick are oversubscribed. Not even an observation requiring a manned moon mission could park there forever. One day bled into the next without a photon to show for it. Had dust kicked up by the Eagle's liftoff shrouded the reflector array? If they didn't find it, Faller knew, NASA would move on, and the Apollo 11 reflectors could well spend eternity as the man in the moon's nostril stud. With time running out, they fired again and again. Finally, on August 1, they captured their first returns from the array.

"We were bloody lucky," Faller says. "Everything worked out perfect that night."

There are now five retroreflector arrays on the moon. Apollo 14 astronauts dropped off one like Apollo 11's in early 1971; the Apollo 15 left a larger one, with three hundred corner cubes, that July; and two long-dead Soviet unmanned rovers, Lunokhod 1 and 2, live on scientifically thanks to French retroreflectors on them. Among other things, these arrays have helped nail down the Earth-moon distance within about a millimeter and established that the moon is moving away from Earth at a rate of 3.8 cm (1.5 inches) a year, that the moon has a liquid core, and that gravity seems to behave like Einstein said it would.[14]

Just shy of fifty-three years after Henry Plotkin's NASA Goddard team lit up the glittering honeycomb beanie Explorer Beacon 22 wore into space, I'm in the control room of MOBLAS-7, Goddard's satellite laser ranging workhorse. It's just up the hill from where Plotkin, Degnan, and the rest of the team did their work, part of what's now called the Goddard Geophysical and Astronomical Observatory, or GGAO. There is a constant zap-tapping at the cadence of a gurgling V-8 engine, corresponding to the laser firing each tenth of a second. Jan McGarry, a NASA Goddard scientist among whose many duties is overseeing GGAO, is my host. She was involved in writing the original software used to aim the laser. She did so in Pittsburgh, where the contractor building the laser-pointing mounts was based. That was in 1977, during the Steelers' Steel Curtain days. Forty years later, despite Maryland roots that should put her firmly in the Redskins' camp, she's still a Steelers fan, and her software subroutines are still helping aim the MOBLAS-7 laser.

Tushar Ujla, a Goddard technologist, is seated at the control station, presiding over a mishmash of flat-screen monitors and Cold War–era beige consoles with push buttons, switches, dials, and red LED displays like the ones on the digital watches of the Steel Curtain era. He's in the middle of his 2:00

p.m. to 10:00 p.m. shift, shooting at satellites. At the moment, he's nailing KOMPSat-5, a Korean radar imager launched in 2013. It's one of about ninety satellites (and counting) with corner cubes, and eleventh on a priority list with sixty-nine entries taped to the console in front of him.

"I'm hitting it pretty good, actually," Ujla says.

The satellite is bright in the center of one of the flat-screen monitor's four partitions. Stars whip past like trees by the highway. The MOBLAS-7 software tracks KOMPSat-5 for a couple of minutes until it's ten degrees above the horizon, at which point Ujla switches to the next target, chosen by a combination of availability and priority. GRACE-A, one of two satellites taking precise measurements of Earth's gravity, is the top priority; Beacon Explorer C, launched a few months after Beacon B, in early 1965, is number 69.

MOBLAS-7 was one of eight MOBLAS systems, the last four built in the late 1970s and early 1980s. These were improvements on the first four NASA mobile systems, from the late 1960s. The biggest improvement was in the laser. Degnan led a team that built a laser with pulse rates of one-tenth of a nanosecond (one hundred picoseconds, or trillionths of a second). Shorter pulses meant sharper distance measurements. That enabled accuracies of about 7.5 centimeters (3 inches). Since then, through various improvements to the detector and timer, they've gotten it down to about one centimeter.[15]

Maceo Blount, a supervisor who will take the 10:00 p.m. to 6:00 a.m. shift in a couple of hours, has come in early to share his insights with me. He has worked with NASA SLR systems since he started at Goddard in 1978. MOBLAS-7 and the ones like it consisted of four trailers, each designed to be driven separately and cabled together at the destination. There was a truck for the radar—it beams out a three-degree cone around the laser system, shutting the laser off automatically if it detects an aircraft. There was a trailer for the lidar and the electronics, which we're in now. There was a trailer for the computers and magnetic tapes and disk platters (it got really warm in there back in the day, McGarry says). And there was a support trailer.

"It was like a circus," McGarry says.

Blount accompanied various MOBLAS lidars to Hawaii's Mount Haleakala, to Mexico, Colorado, California, and Canada. He liked them all, he says, though the black flies got to him in Canada. The system is a dinosaur, he says, but "it's hard to get rid of these dinosaurs because they just keep working."

"They are not extinct," Ujla adds.

**Tushar Ujla operates the controls inside the GGAO's MOBLAS-7 satellite laser ranging station as colleague Maceo Blount looks on.**

Actual mobile SLR stations have, however, gone the way of the dinosaurs. MOBLAS-7 is no longer mobile. Global positioning satellites—American, European, and Russian—combined with thousands of low-cost base stations have provided a cheaper alternative to measuring tectonic plate motion. Very long baseline interferometry (VLBI), a system through which a global network of twelve-meter (thirty-nine-foot) antennas tune into distant quasars, is the principal way of measuring Earth's rotation. There's a French-developed radio-based system called DORIS, in which satellites listen for the transmissions of ground-based beacons and determine their positions based on the Doppler shift. GGAO has all three systems plus GPS running, as do the roughly forty International Laser Ranging Service stations like it around the world.

"There's no point in moving things anymore," McGarry says. "We're looking at essentially the characteristics of the Earth, and we want somewhere that's a really stable location."

They also want a better SLR rig, an effort McGarry is spearheading as well. It's based on the lidar just across the parking lot from MOBLAS-7, in a structure reminiscent of a shipping container but topped with a telescope dome. It was called Next Generation Satellite Laser Ranging, or NGSLR. This, too, was John Degnan's brainchild, though the impetus came from his boss, Goddard scientist David E. Smith, in 1994.

With more and more satellites to track and a need to cut costs, SLR had to automate. Degnan proposed what he called SLR2000, which would replace a high-energy laser firing at low pulse rates with a low-energy laser firing at high pulse rates.[16] To work with a far weaker laser, the detectors would be so sensitive they could register individual photons that had bounced off a satellite, in broad daylight. This became known as single-photon lidar, a technology whose reach would extend well beyond satellite zapping. In addition, the SLR2000 system would automate satellite selection and telescope pointing, sun avoidance, weather monitoring, and more. This new SLR system eventually took the name NGSLR and in 2012 started operating in in experimental mode in tandem with MOBLAS-7.[17]

I've been using the past tense with respect to this NGSLR system despite having seen it with my own eyes. That's because it was felled by a lightning strike in spring 2015. The plan now, McGarry says, is to take its lessons and create a new system capable of measuring the distance to satellites hundreds or thousands of miles away, whipping past at between three and eight times the speed of an assault rifle bullet, to an accuracy of one millimeter or less. Such a system could then proliferate and replace the likes of MOBLAS-7 around the world.

Why bother, if the MOBLAS dinosaur isn't extinct? The big scientific driver, McGarry says, is understanding sea level rise. It's hard to measure the nuances of a global phenomenon that's creeping up just three millimeters a year (some pantry sleuthing found this to be approximately the thickness of a Cheez-It). So you need sharper models of the geoid. And you need to know more exactly the altitudes of satellites measuring sea levels and ice sheets feeding their rise.

"The higher-precision science requirements tend to drive the requirements on the orbits and the need to really have a stable reference frame," McGarry says.

We step outside into the cool October night. Behind us, the massive VLBI antenna swings into action. It moves much more quickly than seems reasonable for something so large, a scientific version of a Steelers tight end.

Atop the MOBLAS-7 trailer, the dinosaur's green pulses flash low into the sky not far above the horizon. I forget to ask McGarry why the laser is green and not some other color. Degnan clarifies this afterward: as SLR took root, neodymium yttrium-aluminum-garnet (Nd:YAG) lasers were found to work better than ruby lasers like the one Plotkin aimed from the missile tracking mount in 1964. Nd:YAG lasers pump out their coherent light in the infrared. But the infrared detectors of the 1970s were inefficient. The solution was to double the frequency from 1,064 nanometers to 532 nanometers, which is doable with some straightforward optics. It's the same emerald green, from the same laser sources, that modern bathymetric lidar uses. Later, I look into what Ujla was shooting at right then. It was a satellite launched in 1976, fulfilling its sole purpose of reflecting laser light back to its source—none other than LAGEOS.

CHAPTER 7

# ZAPPING MARS, MOON, MERCURY, AND EVEN EARTH

As LAGEOS took to the skies, Goddard scientists were thinking about turning the satellite laser ranging approach on its head. SLR stations like the MOBLAS were pricey; corner cubes were cheap. If you could fly the laser and sprinkle corner cubes on Earth, you could grow the scientific bounty and maybe save some dough.

John Degnan was among those thinking about it. So was Jim Abshire, who, like Degnan, had started at Goddard as an undergraduate co-op student, in Abshire's case from the University of Tennessee in 1971. He was soon working on laser communications, the idea being to send data back and forth between satellites. If it could be done, the data rates would be astronomically higher than what radio could deliver. "We had big lamp-pumped lasers and a chiller out in the hallway—the laser had to be water-cooled," Abshire says. We are in his office at Goddard, first thing in the morning. Mugs from various missions—ICESat, MESSENGER, Mars Global Surveyor, and others—are spaced evenly on the windowsill. He looks too young to have been out of preschool in the early 1970s, much less working on space lasers for NASA.

But indeed he was, and indeed he, Degnan, and others thought through the possibilities of lidar in space as they worked on their other projects—multitasking being the norm for a NASA scientist. By the mid-1980s, the idea had crystallized, in the form of what was called the Geoscience Laser Altimetry/Ranging System, or GLARS. It was envisioned as part of a huge satellite called the Earth Observing System. The satellite instrument would have two modes. One, focused on laser ranging, would involve shooting laser light at corner cubes on the ground—SLR with a laser orbiting about three hundred miles overhead. If they were spaced at neat intervals like mission mugs on a windowsill (but thirty kilometers—nineteen miles—apart), you could cover the entire state of California with 157 retroreflectors, they calculated, and target them all every eleven orbits. But you could also put the retroreflectors wherever you liked—on either side of the San Andreas Fault, or near sites where nuclear power plants or pipelines were planned (plate tectonics were a big focus). The second mode, the lidar altimeter, would just zap the ground.[1]

It wasn't to be the first lidar in space. That one flew in 1971, on Apollo 15, and then similar versions rode on Apollo 16 and 17. RCA Aerospace Systems built it, having adapted it from ruby lasers from another RCA division. "They were building tank range finders, which were very rugged," Abshire says. The Apollo instrument was designed initially to fire only when a moon-mapping camera took a photo, the idea being to provide a range estimate accurate to about ten meters, thereby giving perspective to the photos and accuracy to the maps they would inform. But the lidar ran without the mapping camera too; on Apollo 15, it managed to fire about once every twenty seconds (a lag dictated by the amount of time the flashlamps pumping the ruby needed to recover). It failed after four and a half orbits.[2]

GLARS never flew. A later version with its own satellite, called the Geodynamics Laser Ranging System (GLRS), was studied carefully and ultimately dismissed as well. Beyond the technical challenges, an alternative had emerged by the early 1990s: GPS. If you dispersed GPS stations rather than retroreflectors, you could save yourself from building and launching a laser satellite. So that's what they did.[3] Plus, the ground-based corner cubes introduced other complexities: "In team meetings, we were asking how to keep bird shit off the retroreflectors and keep people from shooting at them," one planner said.[4]

But the idea of a mapping lidar in space stuck, and Abshire and his colleagues James Garvin and Jack Bufton got to thinking about proposing an improved moon lidar. The technology had come a long way on all fronts since the Apollo days, in particular with lasers. Rather than flashlamps blasting a jumble of light at some poor ruby rod, the pumping was done by semiconductor diodes dialed to wavelengths known to juice laser crystals, which made things simpler and more efficient and sharply increased the lasers' operating lifetimes. That's right about when David E. Smith came knocking.

It was the late 1980s, and Smith, the LAGEOS progenitor, was heading up a group building a radar instrument called the Mars Observer Radar Altimeter Radiometer, or MORAR. It was to ride on the Mars Observer spacecraft slated for launch in 1992, the goal being to create a topographic map of the Red Planet to an accuracy of ten meters or better.[5] But MORAR kept costing more and more. The Mars Observer project manager finally had enough: MORAR was done. Smith could either find another approach with the money he had left or be thrown off the spacecraft entirely.[6]

Smith knew lidar, and he knew Abshire and colleagues were working on a concept for a moon mission. If they could cobble something together fast and on the cheap, he told them, they could take their idea all the way to Mars.[7]

They formed a team of a few dozen people focusing on different parts of the system: the laser, the detector, the electronics, the software and algorithms, the mechanical, the assembly, the testing, the project management. They couldn't afford to build everything from scratch. They found a flight spare of a spectrometer used for the Voyager missions and adapted it as their half-meter (19.7-inch) receiving telescope. They needed a laser that would sip power, stay cool, and last for months or years. They tapped McDonnell Douglas to adapt one that was originally developed for aircraft smart-bomb target designators, then, later, a military laser communications system that got canceled.

McDonnell Douglas led the Goddard team to a detector that had worked nicely for laser communications: silicon avalanche photodiodes (these trigger an "avalanche" of electrons when a small number of photons strike them). Allan Carswell's old employer RCA had built them in 1985. The detectors were special because they could see the infrared light from mainstay semiconductor lasers, which most silicon materials of the day were blind to.[8] On a single two-inch wafer, RCA had stamped dozens of the 0.7-millimeter dies (0.03-inch, about half the thickness of a potato chip) of military-specification, radiation-hardened integrated circuits destined for the receiving end of an ill-fated lasercom program. They were in storage now. RCA was happy to sell a couple of them to NASA.[9]

Jan McGarry played a big role in developing the onboard algorithms needed to make sense of the light returning to the detector, a job quite like Gary Guenther's with the Airborne Oceanographic Lidar. Mars is just a quarter the size of Earth but is geographically more dramatic—the highest mountain in the solar system is in Olympus Mons, and the longest, deepest canyons are in Valles Marineris. Its clouds of red dust would also bounce back photons well before landfall, and there would be the random noise any detector deals with.

Different topography and surface characteristics also play with reflected pulses. Over the flats, photons coming back might be bunched in nice, tight return pulses; off mountain slopes, they would be longer and sloppier. The software McGarry's algorithms encoded told the system what sort of amplitude and power the photons coming back should have (if it's too low, ignore it) and when to pay attention to it (if it's too soon, it's probably dust, ignore it). Abshire likens the task to looking for a friend in a chaotic scene—a sort of *Where's Waldo?* for space scientists. "The more sensitive you can be as far as where and when to look for a pulse, the more sensitive the ranging," he says.

The Goddard team ultimately got it done, and MOLA, the Mars Orbiter Laser Altimeter, launched with Mars Observer in 1992. About a year later, just three days before it was to pull into orbit at its destination, the spacecraft stopped answering mission control's calls. It was over before it really started.

One response to such a loss would be to wallow in self-pity while lamenting the years of work and millions of dollars down the tubes. But this wasn't NASA's first lost spacecraft, and it wouldn't be the last. Plus, there was good science to be done with a lidar over Mars. The Mars geoid was still a rough draft, and the connections between topography and gravity were a mystery. A lidar could give clues as to past tectonics and volcanism, shedding light on the Martian interior; it could help scientists establish the flows of wind, water, and ice, whose erosional processes depend on topography and which can help establish the age of mountains, valleys, and other features. An accurate enough lidar system could detect changes in the seasonal thickness of polar ice caps. It could establish the slope and roughness of the terrain of possible future landing sites and how much atmosphere future landers would have to plunge through en route.[10]

So they tried again. Abshire brought in Xiaoli Sun, a postdoctoral researcher who had been working on another failed laser communications program, his at Johns Hopkins University. He worked to beef up the electronics surrounding the now decade-old detectors, which boosted their timing accuracy by a factor of four. The laser was improved, the telescope was improved. MOLA-2, as Sun called it, launched with the Mars Global

Surveyor in 1996. By the time it stopped operating ten years later, it had fired 671 million laser shots, input with which Goddard scientist Greg Neumann created the first lidar-based topographic map of a planet. It graced the cover of the prestigious journal *Science*, the tie-dye colors Neumann chose for different elevations popping electrically. MOLA mapped to a vertical accuracy of sixteen inches, which was as close as they could get given the uncertainties of the spacecraft's orbit around something that, to the naked eye from Earth, is a faint orange pinhole in the sky. And indeed, MOLA observed the buildup and evaporation of the polar ice caps, which the initial instrument would have been too blunt a tool to consider.

"It was a huge blow when the spacecraft was lost, but we got to rebuild, and the second one was a far better instrument," Abshire says.

By the time MOLA launched with the Mars Global Surveyor, there had been other space lidar. The Naval Research Laboratory's Clementine spacecraft, a technology tester, orbited the moon in 1994. Its lidar used a detector from that same Canadian batch. Neumann helped out on that program too. He had just arrived from a PhD in marine geophysics from MIT. This might, given the dearth of boating opportunities on the moon, look to be a competency mismatch. But this self-described "child of the sixties" had graduated from Reed College in 1969 and worked as a machinist, an industrial engineer, and a nuclear piping engineer, among other things. He had programmed systems and made measurements and understood how to interpret information arriving in waves of all sorts.

"I knew how things worked," Neumann tells me in a Goddard conference room, to which he has brought colorful maps of rocky orbs and a thick binder. "Spacecraft are no different. Sound waves or light waves, it's all the same. It's just a number."

As a sort of warm-up, he had asked around for the Apollo laser ranging data. They arrived in the form of eighteen emails stacked with four columns of numbers, a few hundred of the three thousand or so soundings made across the three moon missions. His binder is full of pages and pages of Clementine moon data, of which about seventy thousand of the mission's six hundred thousand returns were good enough for him to put together the first global elevation map of the moon.

There were other space lidar missions worth noting. In 1986, a Johns Hopkins Applied Physics Laboratory team launched one of McDonnell

Douglas's lasercom leftovers as part of its Delta 180 test program looking into technologies capable of identifying incoming intercontinental ballistic missile warheads.[11] The year Clementine orbited the moon, the space shuttle *Discovery* carried NASA Langley Research Center's Lidar In-Space Technology Experiment (LITE). This was a two-ton atmospheric lidar, a technology demonstrator to pave the way for future atmospheric lidar-in-space efforts.[12] LITE paved the way for NASA Langley's 2006 joint mission CALIPSO with the French space agency. CALIPSO was still probing the atmosphere in 2018. NASA's 1998 NEAR mission to Asteroid 433 Eros used a lidar with a detector from that same Canadian batch, and Neumann, Smith, and others from the Goddard MOLA team used the data to create a map of the surface of this cosmic sweet potato.[13]

Goddard stayed at the forefront of space lidar. As MOLA-2 headed to Mars, its scientists and engineers used a spare MOLA-2 laser and detector and a bunch of leftover electronics to build a Shuttle Laser Altimeter as part of NASA's Hitchhiker Program, which flew small experiments in the space shuttle bay. The altimeter flew with *Endeavor* in January 1996 and with *Discovery* in August 1997. This was more than a MOLA retread. Most important were more sophisticated (and power-hungry) electronics that hadn't been along for the ride to Mars. They let the "hitchhiker" gather much more information about the laser pulses coming back from the planet than the Mars Global Surveyor or Clementine lidars could. Those data let lidar see more than just the planet's surface: rather than simply spotting Waldo, it could tell what Waldo had in his hands.[14]

MOLA's success on Mars led to a lidar instrument on the MESSENGER mission to Mercury, which launched in 2004. The Goddard team built the laser for that one themselves, but they used detectors from that same Canadian batch. Many of the same players were involved—Jan McGarry did the algorithms, Xiaoli Sun worked with the Canadians to improve the detector, Greg Neumann worked the mapping data. It was MOLA-like, but miniaturized, and it had to deal with the unique challenges of a planet so close to the sun. Lasers are temperature sensitive, and this one was riding a spacecraft that, when crossing from the night side to the day side of Mercury, went from minus 104 to 438 degrees Fahrenheit in the span of five minutes. No earthly test chamber could simulate such a swing. To keep the spacecraft's sunshade blocking the sun at all times, MESSENGER's laser had to fire at the planet's surface from sharp angles, making measurements much more difficult. To reduce its exposure to heat reflecting from Mercury's surface and to preserve fuel, the spacecraft had to settle for a big, elliptical orbit, which exposed the northern half of the planet to the lidar but kept the south beyond its reach.[15]

Among the adaptations were to split up what would have been a single ten-inch receiving telescope into four smaller ones topped with sapphire lenses that were less susceptible than regular glass to expansion and contraction and better able to withstand radiation. While it was, as Sun describes it, "kind of nerve-racking, at least in the beginning," the lidar took forty-two million shots until MESSENGER deliberately crash-landed in 2015.[16] Neumann made his third colorful elevation map of a celestial body, though the frayed transition to the blank white space of its southern hemisphere gave it the impression of an abandoned painting.

Goddard's most recent planetary lidar went back to the moon. The Lunar Orbital Laser Altimeter, or LOLA, launched with the Lunar Reconnaissance Orbiter in 2009. This was a further evolution of MOLA, one distinguished by its splitting of the outgoing laser beam into five separate beams captured by five separate detectors built around those same Canadian specks of silicon. Having the five spots per pulse afforded an instant understanding of the surface slope and roughness across the laser footprints. This information would otherwise have taken multiple orbits to collect. Neumann ended up with billions of data points with which to create yet another colorful topographic map of a heavily cratered body. Another LOLA wrinkle: it could receive green laser pulses from Earth, fired from satellite laser ranging stations that pinged it from four continents to further refine knowledge of the spacecraft's orbit. A team led by Xiaoli Sun used that same capability to advance the science of laser communications. Prior to its 2015 electrocution, Goddard's Next Generation Satellite Laser Ranging System zapped an image of the *Mona Lisa* to the lunar orbiter, which the orbiter then radioed back to Earth. The effort was no more a stunt than Fiocco and Smullin's moon shot. Laser communications could one day solve today's interplanetary data bottlenecks. It worked: Leonardo's masterpiece was grayscale and speckled with errors caused by the Earth's atmosphere, but that smile came through as mysterious as ever.[17]

Many other lidar systems have been to space, and many more are planned. The moon has been a popular destination. Japan sent one there in 2007 on its SELENE mission; China did the same that year on Chang'E; and India's

Chandrayan spacecraft went into orbit with a lidar in 2008. Allan Carswell came out of retirement to work on an Optech-built lidar that touched down with NASA's Mars Polar Lander in 2008. It showed the distribution of dust and clouds and saw snow falling in the Martian sky. Optech also designed a lidar to map the surface of an asteroid called 101955 Bennu as part of NASA's OSIRIS-REx sample return mission, which launched in 2016. A Goddard-built instrument called CATS (Cloud Aerosol Transport System) sent a record two hundred billion pulses into Earth's atmosphere from a perch on the International Space Station. It observed clouds and atmospheric aerosols and monitored dust storms, fires, and volcanic eruptions for two and a half years starting in January 2015 and incorporated the first space version of Ed Eloranta's high spectral resolution lidar.[18]

Lidar is poised to play a growing role in space in the coming years, most pressingly in the study of Earth systems. In January 2018, the National Academies released the Decadal Survey for Earth observation, a road map to help NASA and other agencies decide what flies in the next ten years. The report recommended five major missions, six midrange, and three smaller ones. A lidar mission to measure global clouds and aerosols was among the recommended big-ticket items. Four of the six midrange missions included lidar: satellite measuring of carbon dioxide and methane, of global land and sea ice, of global snow depth and snow-water equivalent, and of the 3D structure of forests and other terrestrial ecosystems. All three of the small missions included lidar. These would measure global winds, track the planetary boundary layer, and map global topography.[19]

The Europeans are moving ahead with lidar as well. After years of delays and cost overruns, the European Space Agency is scheduled to launch Aeolus in 2018. Its ultraviolet Doppler lidar will measure global winds—long a priority for meteorologists and climate modelers.[20] The French and German space agencies are collaborating on MERLIN, a space lidar to measure atmospheric methane, a major greenhouse gas. It's slated to orbit starting in 2021.[21]

Goddard is also in the Earth-sensing lidar mix. Jim Abshire has been working toward a carbon-dioxide-measuring lidar mission for more than a decade. NASA's planned ASCENDS mission, or "Active Sensing of $CO_2$ Emissions over Nights, Days and Seasons," aims to see places that carbon dioxide sensors already orbiting (NASA's OCO-2 and Japan's GOSAT) can't because they use passive imagers that rely on sunlight.

The ASCENDS lidar could observe through the darkness of the polar winter, keeping track of such things as how much carbon dioxide the thawing permafrost is adding to an atmosphere already brimming with more of the greenhouse gas than any time in the last four hundred thousand years.[22] Permafrost locks in about twice as much carbon as the atmosphere hosts at the moment, creating a potentially vicious cycle in which human-caused climate change triggers a natural carbon bomb.[23] ASCENDS will also do a better job in places like the tropics and Amazon rain forests, where sun-reliant satellites tend to be thwarted by clouds (the lidar also has a hard time with thick clouds but can deal with thin ones and can get shots through the gaps, Abshire says).

With ASCENDS, Abshire's team started in the lab and then, exploiting some of the same telecommunications-derived laser technologies NCAR scientist Scott Spuler is using in his atmospheric lidar, built an instrument for an aircraft. The airborne instrument is called the $CO_2$ Atmospheric Sounder. It has flown over California's Central Valley, farmland in Iowa, Nevada desert, forests in the Pacific Northwest, and Alaskan permafrost, among other places. The Alaskan campaign, in 2017, involved fifty hours of flights in NASA's DC-8. The lidar needs to be exceedingly accurate for a space instrument to make sense; Abshire says they're close and that should ASCENDS get the nod, vastly greater budgets for components, calibration, and testing, combined with more tightly controlled operating temperatures in space, should put them over the top.

Abshire recalls a conversation he had with his boss, David E. Smith, as the MOLA project got going thirty years ago. Smith told him if he found an important problem, he could expect to spend five or ten years working on it.

"I was young and thought, 'That's insane,'" Abshire says. "This one's taking longer than that. But it's still important, and it's still really interesting."

We've seen lidar go from Hutchie's mind to the moon, into the air and water, and out to Mars. We've skipped something relevant to a terrestrial species: land, on Earth.

# CHAPTER 8
# LAND AND ICE

To map something from the air or space, you have to know where your aircraft or spacecraft doing the mapping is. Orbital dynamics equations, with help from satellite laser ranging and GPS corrections, could do a good job with satellites. That didn't work for aircraft. The same NASA Wallops Flight Facility team that developed Airborne Oceanographic Lidar (AOL) came up with something that did. Bill Krabill led the effort.

Krabill had worked on the data side of Hongsuk Kim's first iteration of what would become AOL in the mid-1970s. He had grown up a few miles north of NASA Wallops in Pocomoke City, Maryland—the sort of place, as Krabill puts it, where "to be a true native of this particular region, your grandparents have to be born here." He took a summer job at NASA Wallops working data for radar engineers while studying for his math degree at Salisbury University and kept doing it when he graduated in 1966. He worked his way up, commuting in a pickup truck that came in handy on the thirty-acre farm just south of Pocomoke where he lived with his wife, Judy, and, with time, their two kids. He had a beard long before it was trendy. He was, as one colleague puts it, the sort of guy you could picture "around a campfire with a bottle of whiskey, sharing good stories."[1]

Krabill's team worked to advance a raft of technologies to get AOL to do bathymetry. They built more than a bathymeter. As early as 1979, Krabill and colleagues flew the Wolf River Basin east of Memphis, Tennessee, to map the topography of the land and the winding river, even noting the height of trees.[2] By then, they knew their lidar was measuring vertical distances to within about ten centimeters (four inches). What they couldn't tell was what part of the landscape they were so precisely measuring. For that, they had to pinpoint where the airplane was.

Over the Wolf River Basin and elsewhere, they tried about everything. They compared their photos, taken at 225 miles per hour at five hundred to a thousand feet up, with existing aerial photography. They picked out landmarks or parallel roads and used them as guides. They used onboard accelerometers to track the plane's responses to wind gusts or thermals. They filled red and white weather balloons with helium and tethered them so they floated above the trees like buoys, marking a channel for the pilots to fly between. They tried smoke flares; they tried strobe lights. There were problems with them all, says Jim Yungel, who joined Krabill's team right out of college in those years. With the smoke flares, people called in forest fires. Strobe lights attracted the attention of locals, who asked variations on "What are you doing?" and "Can you do that?" And the balloons had the disadvantage of being balloons.

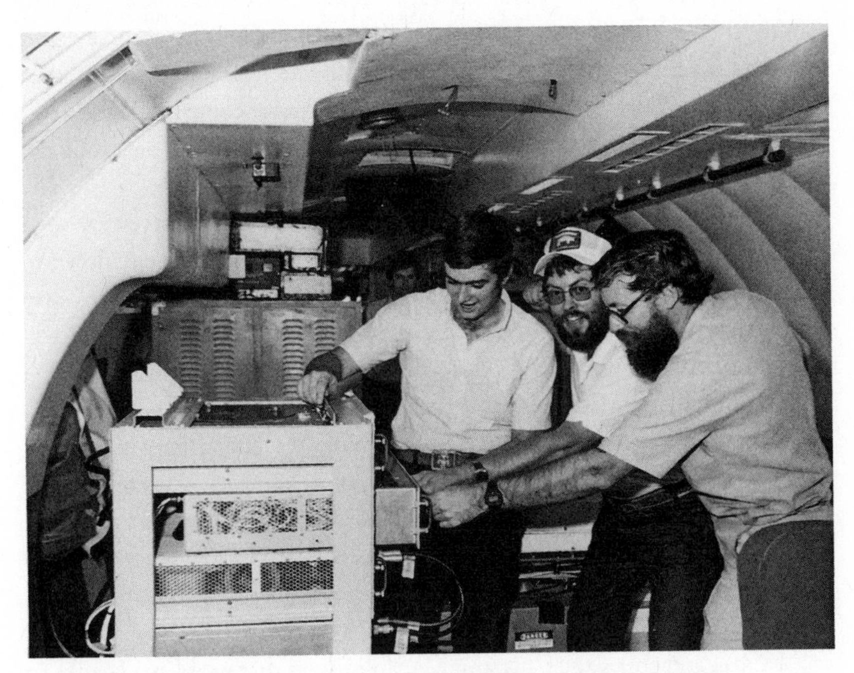

**Jim Yungel (left), Bill Krabill, and Earl Frederick in the NASA Wallops P-3 flying the Airborne Oceanographic Lidar in 1985. (Photo courtesy of NASA/Jim Yungel.)**

"At one point, a pilot says a balloon's moving on him," Yungel says. "A child on a bicycle had cut the string, tied it to the back of his bicycle, and was pedaling away." Even if the balloon didn't bicycle off, reconciling flight lines with the land below made for tedious, time-consuming work.[3]

There were higher-tech approaches such as ground-based radar, but they weren't perfect either. Radio-based systems such as Omega and Loran worked for ships but were only accurate to hundreds of meters at best, far from the range needed for precise airborne land mapping. The US Air Force had started launching GPS satellites, but those looked to be accurate only to within about ten meters, unless you were piloting a fighter jet or other military machine with access to more precise data. Even then, the constellation was incomplete, leaving coverage gaps. Then Krabill came across an article that talked about using GPS in a different way than its military creators had envisioned.

GPS satellites carry atomic clocks, and GPS receivers have access to atomic time; your phone's GPS chip relies on atomic-time accuracy provided by the cell towers. A GPS receiver grabs the time and orbit-location data broadcast from at least four satellites. The receiver compares those times with its own time, then triangulates its location based on how long it took the signals from the satellites to arrive.

The article Krabill came across talked about how surveyors were exploiting something called carrier phase data to get GPS accurate down to a few centimeters. It was the sort of accuracy that AOL would need to map terrain. Carrier phase data exploits the underlying carrier wave's frequency—not just the data the carrier wave delivers.[4] That frequency translates into a wavelength of about twenty centimeters (eight inches). Track the frequency, and that's more or less your positional accuracy. Tracking carrier waves over time could improve it even further. But surveyors had parked nineteen-inch rack-mounted systems in one place for two or three days. Putting GPS in an aircraft was something else entirely.

"They weren't built to do that, and they can't handle the dynamics of an aircraft," Krabill says. "And it was a problem."[5]

Krabill and colleague Chreston Martin tackled that problem. Thanks to its lidar, the aircraft carrying AOL was capable of knowing with more precision than any machine ever flown how far above the ground it was. Krabill and Martin realized that their lidar thereby offered a superaccurate yardstick against which they could compare elevation data from a flying GPS unit.

With this insight, once you had confidence in GPS's vertical accuracy, you could derive horizontal position, which was the big missing piece in airborne terrain mapping. The resulting knowledge of the plane's location, in turn, would open new frontiers for airborne lidar.

On July 31, 1985, the NASA P-3 carrying AOL flew up and down Chincoteague Bay at about four hundred feet altitude with a Texas Instruments GPS unit weighing fifty-three pounds without the antenna. A matching unit stayed on the ground at NASA Wallops. NOAA's National Geodetic Survey, which had provided the GPS units, also chipped in with tide gauges to establish the bay's water level. The team gathered twenty-five minutes of data in the air, landed, and got to work. Among the many corrections Krabill and Martin had to make once they tapped into AOL's various magnetic tapes and two-foot-diameter hard drives were for the aircraft's orientation—a subtle tilt of the nose or the wings could shift the laser's ground spot many feet, affecting the lidar's measurement. Also, the GPS antenna was mounted to the top of the aircraft and forward of the laser firing from below it. Those offsets in turn exacerbated the effect of the two-degree upward pitch of the flying P-3. It all had to be established with millimeter precision. They reconciled the laser's measurement to the water's calm surface with the GPS unit's measurement to a smooth, theoretical ellipsoid about sixty-six feet below it. When they were done, the results matched up within about twelve centimeters (almost five inches), and Krabill and Martin thought that with a full GPS constellation and better hardware, they could get it down to a centimeter or two.[6]

Krabill scheduled summer meetings at NASA headquarters in Washington, DC, in those days. These were, he says, "to give a quick dog-and-pony show about what we were up to and to get folks interested." He brought his wife and kids along. They went to the Smithsonian museums and saw the sights and made a family vacation of it.

Not long after that GPS flight over Chincoteague Bay, Krabill presented the results to a group of NASA science program managers. One of those present was Robert H. Thomas, who had recently joined NASA from Ohio State University. Thomas was impressed with the accuracy in elevation that lidar could deliver, and also that the aircraft could go where a human pilot wanted it to, not just where Newton's laws said it had to go. "He said, 'If you can do this, you can do some good stuff for ice science,'" Krabill recalls.[7]

Bob Thomas knew his ice. Thomas, a Brit, had spent 1960–1961 deployed as a meteorological observer at Faraday Station in Antarctica while in his early twenties. He got interested in ice caps—not in simply mapping them, as was the main thrust at the time, but in examining how they were moving. He spent his free time learning to do ice sheet surveys, which involved strain gauges and tape measures and driving in stakes and watching them move with surveying equipment. That led to a second visit to Antarctica five years hence, this time, as he puts it, "measuring the hell out of the Brunt Ice Shelf and its adjoining bit of ice sheet" with those same tools, with which he skidded about on a dogsled. He realized that the Brunt Ice Shelf was acting like a cork, holding back the grounded glacier behind it. He coined the term "ice shelf buttressing," demonstrating the concept during his talks by uncorking a bottle of wine and pouring a glass. The point was simple: if the ice shelves surrounding West Antarctica melted, there was nothing to hold the glaciers behind them back.[8]

In the late 1970s, whether or not ice sheets in West Antarctica were losing mass (that is, melting more than they were accumulating) wasn't clear. What would happen with more warming was. Consider the title of a 1978 article the glaciologist John Mercer published in *Nature*: "West Antarctic Ice Sheet and $CO_2$ Greenhouse Effect: A Threat of Disaster."[9] Mercer's estimate of a five-meter (sixteen-foot) sea level rise with West Antarctica's global-warming-triggered slip-and-slide into the sea was overstated: modern revisions peg the potential rise at 10.8 feet—not much higher than a basketball hoop.[10] Thomas followed up in 1979 with his own report in *Nature*, supporting Mercer's conclusions and adding that while a complete removal of ice shelves would bring about a collapse of the ice sheet in less than a century, a "rather slow process" taking several hundred years was more likely. Four decades later, his conclusion holds.[11]

Thomas's interest in polar ice sheets never waned, and his preferred tools became more sophisticated than the assemblage with which he once dogsledded around the Brunt Ice Shelf. One option was satellite radar altimeters. These were what they sound like: satellites blasting radar pulses at the planet from orbit and measuring the time it took for the microwaves to bounce from the surface to the spacecraft's antenna. Given the long wavelengths and fat beam angles of radar, the altimeters' footprints measured well over a mile across, and the raw data tended to be noisy. Microwave altimeters were good at measuring big, flat areas, and they became a mainstay for monitoring oceans.

Thomas wondered if radar altimeters would work on ice sheets, which are sometimes flat and sometimes anything but. He did a lot of work with H. Jay Zwally, a physicist who had become an accidental glaciologist in 1972. That's when funding for research in physics had nosedived, and Zwally, facing unemployment at the University of Maryland, took a job as a program manager at the National Science Foundation and was soon managing the glaciology and remote sensing program. He moved to NASA Goddard in 1974 to begin his research in polar glaciology and sea ice. While establishing Goddard's cryospheric research program, he also focused on wringing information about ice sheets out of satellite radar altimeters starting with the very first one, GEOS-3, launched in 1975.[12] Zwally and colleagues managed to estimate the surface elevation of the portion of southern Greenland from GEOS-3 to an accuracy of about two meters.[13] A couple of years later, Zwally, Thomas, and others took data from the short-lived Seasat radar altimeter and mapped some of Greenland's as well as Antarctica's ice sheets with improved accuracy.[14]

If Thomas's principal interest was the movement of ice shelves and the glaciers they bottle up, Zwally's was mass balance. That is, he wanted to know if the ice sheets were growing or shrinking. Radar altimetry's low resolution and the problems with slopes of more than a degree or two made that hard to do.[15] The answer, they both recognized, would be in radar's optical spawn. In 1981, they collaborated on a paper with NASA Goddard glaciologist Robert Bindschadler, proposing a satellite laser altimeter system whose seventy-meter-diameter (230 feet) spots of laser light would, over time, dot nearly the entirety of the poles, enabling measurements to an accuracy of about ten centimeters (four inches) and measuring topography in ways radar just couldn't.[16] In terms of precision, satellite radar altimeters were a house-painter's roller. Laser altimetry would be an artist's paintbrush.

All this is to say that as Bill Krabill talked about the potential of the lidar-GPS combination to do unimaginably accurate aerial elevation mapping over large swaths of land, fast, Bob Thomas was primed to see its potential in chillier places. Thomas provided more than just encouragement. He steered money to Krabill's Wallops team to further develop GPS integration with the aircraft's inertial navigation system to pave the way for work over ice.[17]

In 1991, they flew AOL over Greenland for the first time, scanning 500-foot-wide swaths from an altitude of 1,500 feet. The aim wasn't to bring home geodetic-quality data, Krabill says, but to continue to prove out the system. The GPS constellation had only fourteen of the twenty-four satellites needed for round-the-clock coverage, and their orbits were such that the AOL team could only use the constellation to locate their P-3 about four hours a day. But by 1993, when a flurry of launches filled out the GPS roster, they had refined the system so they could establish ice sheet heights to about ten centimeters. Thomas had pushed Krabill's team to get to that accuracy, which was enough to be able to see year-to-year changes in the ice.

AOL flew for hours on end. Krabill sat up front, describing the coasts as incredibly beautiful and the ice sheets themselves as incredibly boring. "The best missions were the ones where'd you'd walk up and down the aisle and half the people back there were sound asleep," he says. "You knew everything was working fine."

Exact GPS positioning became even more critical as the years passed and they re-flew flight paths to see what had changed. If pilots had gotten distracted or flown the wrong fjords on those early flights, they flew them again anyway. "The glaciers that the pilots flew by mistake turned out to be very valuable data," Yungel says.[18] By 2000, the team had enough data to report that the 1.2-mile-thick ice sheets in Greenland's center appeared to be holding steady, but that consistent calving and melting at the fringes contributed to about 7 percent of observed sea level rise.[19] Waleed Abdalati, then a postdoctoral researcher and later NASA's chief scientist, was part of Krabill's team in the late 1990s. The findings, he says, were key evidence that "Greenland was starting to really stir."[20]

Over time, AOL evolved into the Airborne Topographic Mapper. In the years hence, ATM has flown in different aircraft over Greenland, the Arctic, and Antarctica, taking measurements not only of ice sheets but also of sea ice, which has been in stark retreat since the days Krabill's team first flew.

Meanwhile, Jay Zwally kept pushing for a lidar to measure polar regions from space. One would have flown with a polar scientist—ideally, him—on the space shuttle's first polar flight in the late 1980s. He had won NASA Goddard leadership's support for it by January 1986. But the *Challenger* disaster struck two weeks later, and hopes of Zwally's spaceflight ended there.[21]

Recall, though, the big Earth Observing System satellite that was to carry all sorts of instruments, including the GLARS satellite laser ranger. It lived on, in a way, though as a bunch of smaller missions. (If you search online for "Earth Observing System," you'll find that the original satellite evolved into a long-standing NASA program involving many satellites.) Breaking things up made sense: a spacecraft measuring polar ice sheets needs a different orbit than one studying tropical rainfall. And from an engineering perspective, all the interdependencies of a big spacecraft adorned with many instruments drive costs up to the point that it's cheaper, despite launch costs, just to keep things simple. GLARS and its follow-on GLRS, which Zwally was deeply involved with, would both have included an ice-sheet-measuring laser altimeter. That piece of GLRS survived and became GLAS, for Geoscience Laser Altimeter System, the lidar at the heart of ICESat (Ice, Cloud, and Land Elevation Satellite).

Zwally became ICESat's project scientist. He enlisted Jim Abshire as the instrument lead, who in turn brought in Xiaoli Sun, Jan McGarry, and others who had worked on the Mars MOLA lidar and its progeny. ICESat's GLAS lidar had three infrared lasers evolved from MOLA for altimetry (the tripling up was for redundancy and to extend mission life) and used those same Canadian detectors. ICESat also carried a lidar based on the green, photon-counting approach John Degnan had invented for satellite laser ranging. It would probe the atmosphere.[22] The altimeter's 230-foot-wide laser footprints were exactly as wide as Zwally and Thomas had proposed in 1981, and its forty pulses per second would dot the ice every 558 feet.

ICESat launched in 2003 and, despite laser problems that required scientists to do a succession of intermittent, thirty-three-day campaigns, managed to operate until 2009, when its last infrared laser petered out. It had by then fired nearly two billion shots, with a vertical accuracy that, with refinements, narrowed to just two centimeters (three-quarters of an inch).

The spacecraft answered Zwally's forty-year-old question of what was happening with the mass balance of ice sheets. The data showed that despite net losses of ice in West Antarctica and the Antarctic Peninsula, snowfall elsewhere, particularly in East Antarctica, led to an increase in the mass of Antarctic ice large enough to counteract global sea level rise by 0.23 millimeters (0.009 inches) a year. Zwally and colleagues added, though, that increased West Antarctic melting of the sort Bob Thomas warned about could overtake the additions of East Antarctica within about two decades without increases in snowfall accompanying a warmer climate (such increases being a possibility, as humidity will rise globally).[23] Over Greenland, on the other hand, ICESat saw annual melting speeding up, with faster discharge around the flanks and slower accumulation high on the ice sheet, resulting in a global sea level rise contribution of about a half millimeter (0.02 inches) a year—that's 17 percent of the three-millimeter (0.12-inch) total, more than twice as much as Krabill's team had noted a decade earlier.[24]

ICESat data found that Arctic sea ice's freeboard (the tip of the iceberg, so to speak) height fell an average of about 1.6 centimeters (0.63 inches) a year from 2003–2008 and that the coverage of multiyear ice (ice that survives the summer) shrank 42 percent during that same period. The ice that didn't melt thinned by about two feet.[25] That's critical information, given that the disappearance of Arctic sea ice means dark ocean waters absorb more sunlight, speeding up warming in a region that has seen temperatures rise faster than anywhere else on Earth.[26]

Even as ICESat orbited, plans were afoot for its replacement, ICESat-2, slated for launch in September 2018. The original idea was to launch a copy of the first ICESat. That turned into a more ambitious mission, one with six infrared laser altimeters as workhorses and an experimental system using green, photon-counting lasers of the sort Degnan invented for comparison. When that looked too expensive, NASA discarded the infrared and went solely with the green.[27] Rather than firing single pulses forty times per second as the original ICESat did, ICESat-2's laser will split into six beams, each pulsing a forty-foot spot on the planet's surface ten thousand times a second. Despite orbiting along its 310-mile-high path at seventeen thousand miles per hour, that pulse rate will flash the center of each new spot just twenty-eight inches past that of the previous one. The six beams will flare out in three pairs, with ninety meters between each pair and six kilometers (four miles) across the lot of them, providing instant slope information much like LOLA, Goddard's lunar lidar. All told, ICESat-2's vertical accuracy should be about four millimeters (0.16 inches), give or take. So pretty incredible, and enabled to no small degree by photon-counting lidar, which can do it all with a seventy-fifth of the laser power of the original ICESat. All this innovation has come at a cost, however: problems with the laser delayed launch from 2015 to 2018, and ICESat-2 is now slated to set US taxpayers back $1.1 billion.[28]

As soon as the first ICESat's lasers started failing too soon, it was clear there would be a gap between the first and second ICESats; now it will be, best case, nine years. Radar and gravity satellites are filling in, and so is the NASA Wallops team that Bill Krabill first flew over Greenland in 1991. The ATM team's main mission today is Operation IceBridge. They re-fly their old routes over Greenland, the Arctic Ocean, and Antarctica and add new ones. The latest version of ATM pulses ten thousand times a second and includes both infrared and green lasers—the first for continuity with the original ICESat, the second to pave the way for its successor. Krabill is retired now, but his son Kyle has joined the team as a computer expert and keeps him up to date. Both see the need for the next ICESat, particularly on the southern continent.

"You can get your arms around Greenland in a P-3," Krabill says. "You can't really get your arms around Antarctica."

NASA Wallops researchers led the way to modern lidar bathymetry and, thanks to its pioneering work with airborne GPS, topographic mapping. Krabill's chance meeting with Bob Thomas at NASA headquarters led to a new mission monitoring ice sheets. Along the way, Krabill opened the door to yet another use for lidar. It had to do with forests, both with the trees and what they were hiding.

# CHAPTER 9
# TREES AND ARCHAEOLOGICAL TREASURES

During those 1979 flights in the Wolf River Basin west of Memphis, the NASA Wallops Airborne Oceanographic Lidar team was all too aware of the trees. Bill Krabill and colleagues flew in the early spring before the deciduous ones went green. That was intentional. Leaves would block the laser and might even thwart airborne topographic mapping efforts entirely for several months a year in midlatitudes, they worried.[1]

But that same year, Russian researchers had used an airborne lidar expressly to check out profiles of tree heights.[2] There were good economic and ecological reasons to try lidar in forests. Surveying forests involved the cumbersome work of hiking into the woods to measure trunks and estimate heights. Companies in the lumber, paper, and other industries might use lidar to quickly assess the state of their inventory. Similarly, ecologists knew that a forest's vertical and horizontal structure can say a lot about the age of a forest, the types of trees involved, and the sorts of habitats it provides. Forest managers were interested in better ways of assessing forest health as well as fire danger. And while it wasn't top of mind back then, the role of forests is one of the big unknowns of the global carbon cycle: if the world's biomass converts less carbon dioxide into wood, temperatures could rise faster; if it converts more, it could buy us more time.[3]

Krabill and colleagues published their Wolf River Basin results, as they did with all their notable efforts. Gordon Maclean, a master's student in forestry at the University of Wisconsin, came across the paper. Maclean's thesis work, which he was wrapping up, centered on resurrecting an idea published by German scientist Reinhard Hugershoff in 1919 and applying it to modern forestry. Hugershoff was an aerial photogrammetry pioneer, and he had advanced the technique to determine how big things were based on two or more photographs taken from different angles. Maclean's thesis used the approach to estimate the number and size of trees in a forest.[4]

Krabill and colleagues had seen trees as obstacles blocking the terrain he was trying to map; Maclean saw them as the Russians had (though Maclean hadn't been aware of the Russian effort, which was published only in Russian). He sent Krabill a copy of his master's thesis.[5]

It didn't take much convincing that AOL's potential in the woods should be further explored. Maclean hired into EG&G, a key contractor at NASA Wallops. AOL had flown over the forests of Doubling Gap in Pennsylvania before Maclean got to Wallops and, soon after his arrival, flew over forests near Snow Hill on Maryland's eastern shore. The lidar and the photogrammetric results matched to within less than a meter.[6] Later, AOL flew over International Paper Corporation forests in southwestern Georgia, and shortly thereafter added trees to a combined topography-bathymetry mission at the Savannah River nuclear site in South Carolina. Here, the US Department of Energy's main interest was in doing bathymetry in the site's cooling and effluent ponds as well as topography along the creek that drained them to the Savannah River.[7] This being six weeks before Krabill and Martin's GPS test flight back at Wallops, they used the whole arsenal of weather balloons, ground tarps, and so forth to keep the pilots in the right neighborhood.

Maclean flew along in the P-3. Once they got into the air, he observed as five guys took several minutes to boot up the equipment and tap away at thick keyboards, eyes on five-inch green-phosphorous screens. The turboprop's racket was such that "it was hard to hold a conversation when you were standing right next to somebody." He went up front at one point to check out the cockpit. It felt like the treetops were tickling the tail-heavy sub hunter's aluminum belly. One of the pilots remarked, "You realize if the engines cut off, we have about two seconds of glide time."[8]

Maclean returned to the Savannah River plant that October, this time on the ground to count longleaf pine and water oak and assess their heights and canopies.[9] He recognized that these forest lidar flights had produced data worthy of a PhD. He returned to Wisconsin, where he earned one in 1987. Years later, his advisor told him his dissertation committee recognized that the work was significant and unique. But, the professor added, "it was such a new thing to them that they had no real clue of how it fit into the whole framework of remote sensing."

Maclean was a decade ahead of his time. Even AOL, the finest and most versatile airborne lidar the world had ever seen, wasn't quite up to the task of forest surveys. That had to wait for something called full-waveform lidar.

**An extremely simplified example of airborne/spaceborne full-waveform lidar data from a lone Christmas tree. The vertical axis at left represents the laser light's travel time, which translates into distance. The squiggly line depicts the return energy based on the number of photons returning to the lidar—the more foliage at a given height, the more photons bounce back from that height.**

A standard direct-detection lidar shoots a pulse and waits for it to come back, noting the time when it does. But what exactly "time of arrival" means is open to interpretation. The billions of photons in an eight-nanosecond laser pulse extend eight feet through space on the way out; depending on how many things those photons hit on the way down, it can take those photons a lot longer to wander back. People like Jan McGarry programmed the Mars

MOLA lidar to pay attention to the densest cluster of returning photons rather than simply the earliest arrivals. The strongest part of the return pulse was probably from the actual Martian surface (as opposed to a dust cloud hundreds of meters up or a boulder on the ground). McGarry and her NASA Goddard colleagues used the "shape" of the pulse, defined by how many photons came back when, to make assumptions about the roughness and slope of the landscape, among other things.

Full-waveform lidar takes that approach a step further. As the returns from a pulse come back and the detector converts the thumps of photons into electrons, speedy detector electronics sweep those electrons into buckets of time just a fraction of the duration of the original pulse. Consider a full-waveform pulse descending upon a lone Christmas tree in a field on a crystal clear winter night. The pulse would first bounce off the star on top, filling the initial time bucket; successive buckets would fill with more returns as more photons bounced off the snowy branches and ornaments and needles of the thickening lower levels. A sensitive enough lidar might pick up faint returns from the meadow's grasses. Then there would be a final, larger return from the snowy ground. Full-waveform lidar captures it all via the varying number of photons that end up, after translation into electrons, in those different buckets based on slight differences in arrival time.[10]

The NASA Wallops AOL team dabbled in full-waveform lidar to do ocean chlorophyll fluorosensing experiments as early as 1986.[11] But NASA Goddard became the nexus of full-waveform lidar development.[12] The Shuttle Lidar Altimeter, that "Hitchhiker" on *Discovery* and *Endeavor* in 1996 and 1997, was a full-waveform lidar. It had been developed at about the same time as an airborne full-waveform lidar called ATLAS, which Goddard scientist James Garvin and engineer Jack Bufton built.[13] Bufton's colleagues Bryan Blair and David Harding added a scanning mirror and the ability to split the outgoing beam into five cross-track footprints (i.e., all side by side, perpendicular to the direction of flight) and rechristened it SLICER, for Scanning Lidar Imager of Canopies by Echo Recovery.

Harding was a PhD geologist who had started as a postdoc at Goddard in 1988. He was keenly interested in the potential of lidar in his field and beyond, and he had a knack for recognizing how the things the laser scientists were building might be of use to geologists, ecologists, glaciologists, and other scientists looking for better tools. In the spring of 1994, Harding was giving a talk about SLICER at Goddard that included full-waveform lidar returns from a forest. A postdoctoral researcher there, John Weishampel, took interest.

Weishampel's PhD work had focused on using a combination of radar data and computer models to see a forest's structure and how it changes over time. The aim was to understand how forests store and release carbon into the atmosphere. Trees lock up carbon when they grow and then release it back into the atmosphere when they die and rot. The makeup of a forest—the sizes, types, and ages of trees—has a lot to say about its carbon balance over time. Carbon balance is highly relevant to policy as the world warms.[14]

Remote sensing with radar or passive imagers with wide fields of view had a serious weakness when it came to forests. The distribution of big, medium, and small trees averaged over large areas doesn't hold when you look at smaller plots. In particular, big trees and smaller ones tend to congregate because the big ones shade out midsized ones. Hank Shugart, a University of Virginia forest ecologist who had been Weishampel's PhD advisor, likens it to a playground with a few high school kids and a bunch of kindergartners. "The high school kids don't really give a hoot what the kindergarten kids are doing there anyway," he says. The upshot, he adds, is that "the average over a landscape does not typically occur with any frequency at all at any single point in a landscape. Which sort of means that if you sample things with big sensors, you don't necessarily know what the hell's going on the scale of a small plot, which is how trees interact with each other."[15]

Lidar's small footprint and its direct representation of forest canopy structure (radar demands a lot of postprocessing to distill usable data from forests) might solve that problem, Weishampel saw. He told Shugart about it and sent him an image of a lidar trace he got from Harding. Shugart showed the trace to a new PhD student he was advising, Michael Lefsky.

Lefsky had studied wetland ecology; worked for a consulting firm in Washington, DC; and decided to go back for a PhD at the University of Virginia. "I had been thinking about remote sensing in forests—particularly the structural part, like the number and size of stems in forests," Lefsky tells me in his office at Colorado State University, where he is a professor now. When Shugart showed him the lidar trace from Harding via Weishampel, "I looked at it and I said, 'OK, that's what I'm going to do.'"

Harding, as Lefsky puts it, was "generous enough to fly over forests in

Maryland and North Carolina" so he could analyze the data upon which he would base his PhD. The work involved not only going out and measuring trees in research plots but also unpacking the jumble of information squeezed into the squiggles of thousands of lidar returns. In a single return pulse covering a diameter of several meters (the case with SLICER), the lidar trace would capture more than one tree, plus whatever was in the vegetation understory and the underlying terrain. All that would be influenced by the reflectivity of everything in the pulse, in addition to, potentially, the slope and makeup of the ground. Lefsky was doing with forests what Gary Guenther had done for water with the Airborne Oceanographic Lidar's data fifteen years earlier. It was hard, and it was enormously time-consuming.

"I've always said I think my first wife knew that she would have to divorce me when I was sitting with those topographic maps on a family vacation and she was like, 'What are you doing?'" Lefsky says.

Collaborating with Harding and local remote sensing experts such as Warren Cohen, he focused on forests in five locations across Washington and Oregon. These were the sorts of dense stands that, while covering only a quarter of the Earth's land, hold about 90 percent of all terrestrial biomass.[16] Harding's principal interest was in algorithms capable of cutting through all the noise and confusion to reconcile ground truth with what the airborne lidar was seeing, a prerequisite for any decent remote sensing system. His ultimate aim was to be able to characterize forests and the land underneath them from space.

That nearly happened with the Vegetation Canopy Lidar mission, a low-cost Goddard-built satellite led by University of Maryland geographer Ralph Dubayah. This was a spaceborne version incorporating aspects of SLICER and a follow-on airborne sensor called LVIS.[17] It was harder to do in space, and instrument and other problems led to the mission's cancellation in 2000 after three years of work.

Lefsky's collaboration with Harding would pave the way for forest-sensing lidar from a different satellite. Out of what you might call strategic curiosity, Harding flew SLICER higher than usual to take data with a fatter footprint, one more like that of the lidar on the forthcoming ICESat, which would also have a full-waveform detector. Lefsky showed that ICESat would be able to make sense of the trees despite taking in a lot of forest per pulse. He landed on the mission's science team. Using ICESat data as the primary source, Lefsky and a group of international collaborators mapped the carbon stock of tropical forests across seventy-five countries around the world. They produced a global forest biomass estimate spanning nearly 9.5 million square miles—about three times the size of the lower forty-eight US states—and provided much-needed specifics on the state of forest biomass on a country-by-country basis, for many of which data were scant. Initiatives such as the United Nations' REDD program, through which wealthy countries pay developing ones to stop clear-cutting and slash-and-burn farming, use it as a baseline.[18]

A more recent paper explains why we should care: 35 percent of pre-industrial forest cover is gone. Of what remains, as much as 82 percent is degraded from logging, urbanization, agriculture, and such. In addition to contributing to climate change (cutting or weakening forests releases aboveground and belowground carbon), forest degradation can change weather patterns, cut biodiversity, limit water resources, increase fire risk, and more.[19]

Lefsky also worked on the team defining what science ICESat-2 could be expected to bring home. He concluded that it would deliver about the same performance as the first ICESat's when it comes to forests.[20] He wasn't enthusiastic about spending a decade analyzing data no better than those he had analyzed a decade earlier. "Just for me personally, it wasn't worth it," he says.[21]

Another space instrument is on the horizon that will be more interesting to the likes of Lefsky. This one will, like the CATS atmospheric lidar, hang off the International Space Station's Japanese Experiment Module—Exposed Facility. It's called GEDI, for Global Ecosystem Dynamics Investigation, and it's scheduled to launch in late 2018. Ralph Dubayah, the University of Maryland professor who presided over the doomed Vegetation Canopy Lidar mission, is leading it. GEDI is half of another canceled mission called DESDynI, which was to combine a full-waveform lidar with radar, harnessing the strengths of each.[22] NASA pulled the plug on DESDynI in 2011.[23] But it's being reborn across two missions: a NASA–Indian Space Research Organization collaboration called NISAR is doing the radar, and Goddard is building the lidar.

When I visit GEDI in October 2017, it's in a state of undress in Building 11. Jim Pontius, the program manager, points out its refrigerator-sized box of aluminum exoskeleton. It's honeycombed and gold from the iridite antioxidant coating, hanging horizontally like a pig on a spit between the posts of a burly white rack that looks capable of supporting some huge multiple of the

1,300 pounds GEDI will weigh once filled.

GEDI's three lasers will, with help from piezoelectric crystals (which convert motion to electricity, and vice versa) to dither their beams, send a set of eight infrared pulses toward the planet 242 times per second. The eight ground spots, each twenty-five meters in diameter, will span a 2.6-mile-wide swath perpendicular to the ISS's flight path. The detectors absorbing those photons are built around those very same thirty-plus-year-old Canadian semiconductor dies whose siblings orbited Mars, the moon, Mercury, and Earth (in ICESat). GEDI is using the last of them to enable the translation of forest height and canopy measurements into estimates of aboveground carbon, how it's changing, and how forests may or may not be able to sequester it in the future.

So far, things seem to be going to plan, though there's been some improvisation. It started with Building 11, which is usually a propulsion engineering lab.

"I made a deal with the propulsion branch," says Pontius, whose sports sunglasses rest on a clean-shaven head. He's in his early fifties and fit enough that he is nursing a soccer injury. Part of the deal was to build and then leave behind a Class 10,000 clean room, separated from the rest of the high bay via floor-to-ceiling translucent plastic sheeting.[24]

GEDI is cheap, Pontius says. To use "cheap" to describe something that costs $90-some million, you have to be in the space industry. But when you're making something capable of, as Pontius puts it, "decadal science" for the price of a four-year deal for a decent NFL quarterback, "cheap" applies.

They're able to do it for that price for a few reasons. For one thing, GEDI doesn't need its own spacecraft. For another, it doesn't need its own rocket (SpaceX will launch it in an unpressurized "trunk" below its Dragon capsule on a space station resupply mission). But even though the space station will supply power, coolant, data connectivity, and communication with the ground, there have been surprises. One is that the ISS can get stuck in orbital ruts, overflying the same path repeatedly. That's bad if you're an instrument with a two-year life span (a Japanese refrigerator of an instrument is slated to replace GEDI, even if it's still healthy and gathering epochal data) trying to cover as much territory as possible. So they had to add a gimbal—a mechanical wrist—capable of rotating the entire box 6.5 degrees in either direction to target fresh forest. The gimbal motor was a flight spare from an earlier mission, as it would have been too costly to build a new one. Pontius and colleagues also learned that the ISS orbits at a 3.5-degree tilt, as if doing a slight wheelie. That meant the optics inside GEDI had to be mounted at an opposite 3.5-degree angle to compensate so that the lasers fire straight down.

"Man, you would not believe how that complicates things," Pontius says. Through the plastic sheeting, two technicians in white bunny suits and blue latex gloves pick at a plastic box full of a jumble of stainless steel tubes bent at strange angles. The tubes will connect up and snake coolant through the maze of hardware to be crammed into GEDI.

Perhaps 140 people will have worked on this elaborate space lidar at Goddard alone. I wonder aloud if when GEDI surrenders its seat to its Japanese successor, they'll be able to visit their creation at the Smithsonian, which is where expired Hubble Space Telescope instruments often land. No, Pontius says: GEDI will be jettisoned, burn, and land in an ocean far from the forests it will have measured better than any space instrument before it.

Jim Downie's interest was far removed from those of Lefsky and Pontius. It was 2006, and Downie was director of Vegetation Management for Xcel Energy, which had 3.5 million electricity customers in eight states. Trees and power lines don't mix. With seventeen thousand miles of transmission lines (electron highways) and seventy-two thousand miles of distribution lines (electron side streets), Downie had his work cut out for him.[25]

The problem of trees and power lines was most starkly illustrated the afternoon of August 14, 2003. On a hundred-degree day in Ohio, overgrown trees and major transmission lines mixed five times over the span of two hours, tripping the big power lines.[26] Those trees and the deeper problems they exposed pulled the plug on the homes of fifty million people in the eastern United States and Canada.

"Vegetation is the largest single cause of electrical service interruptions on the electric distribution system across this country," Downie says. "Always has been, likely always will be."[27]

In the wake of that blackout, the word "Council" in "North American Electric Reliability Council" was changed to "Corporation," and the organization's power grew commensurately. In 2006, NERC was working on vegetation management standards that they'd release the following year.[28] Downie was looking for better ways to avoid tree-related outages. At the International

Society of Arboriculture's annual conference in 2006, he was giving a presentation on the causes of such outages. They fall into two broad categories: mechanical, where trees knock down poles and wires, and electrical, where there's a short-circuit from a tree touching a wire. Either way, you can end up with a wildfire. Wildfires sparked by trees and power lines have killed many and cost billions. One of many examples is the Wine Country fires of 2017. Sonoma County in California is suing the utility PG&E for damages associated with the fires, which killed forty-five and did an estimated $10 billion in damage.[29]

A guy with a high-end British accent approached Downie after the presentation and asked if he could show him some technology. Sure, Downie said. Alastair Jenkins was a physicist and optical engineer who was on his second lidar-centric startup, this one called GeoDigital.[30] He had, earlier in his career, collaborated with Allan Carswell and Optech. By the late 1990s, Optech was selling ALTM scanners complete with software to create models of wires and calculate the locations of "danger trees" to utilities in the United States, Japan, Russia, and South Africa.[31] Jenkins opened his laptop. "This is lidar," he said, walking Downie through some data. "Do you see any application in the utility vegetation management arena?"[32]

Downie did. He launched a program in which Downie's Vegetation Management team and Jenkins' GeoDigital people developed a system to scan transmission corridors from a helicopter using a lidar made by the Austrian company Riegl. They started with a test along a roughly thirty-mile section of power lines in Minnesota in 2007. By 2009, with the mountain pine beetle die-off in full swing in Xcel's territory in Colorado, they were flying several hundred miles a year, fusing lidar, digital imagery, and forest-specific data to not only precisely locate and size up a tree but also determine whether a tree was healthy, sickly, or dead. They could set virtual distance limits, highlight risky trees, and even simulate toppling ones that might threaten a power line. Crews with tablet computers could access the data in the field to identify offending trees and document their removal. By 2014, Xcel had cut back 250,000 trees through the program, saving millions of dollars in improved efficiency alone.

Downie and his team expanded the effort to take wildfires into account. Power lines can spark fires, but electrical infrastructure is also vulnerable to fires started elsewhere, causing outages and costing a lot to fix, especially in remote, rugged terrain. Downie had a fire scientist estimate forest density and fuel volume at scores of sites and then account for the type of utility pole or structure and its materials (which can be wood, steel, or aluminum). Downie's team used lidar data and imagery to do the same thing and compared the results. They largely jibed, which meant they could now assess the fire damage risk to hundreds of structures with the push of a button. That saved time and money while improving the fire resilience of key transmission infrastructure, Downie says.

Downie has since moved on from Xcel to spread the gospel of lidar and other remote sensing technologies to the utilities industry with a group called Environmental Consultants. Two GeoDigital execs, Scott Rogers and Chuck Anderson, joined him. Anderson estimates that 70 percent of utilities are using lidar to manage vegetation now. In addition to "traditional" helicopter-based and emerging unmanned aerial systems, mobile mapping shows real promise for power companies. Eighty percent of a utility's vegetation management expense is local—the equipment and wires webbing throughout cities, towns and neighborhoods. Lidar-plus-camera rigs on roving vehicles can see not only the trees, poles, transformers, and wires with millimeter accuracy, but also other elements of a utility's vast, complex infrastructure, Rogers says. That sort of information can help a power company manage things more efficiently, making for a better, less costly system that benefits us all.[33]

All this talk about trees aside, Bill Krabill's original sense that they were just in the way persisted. Two scientific fields would benefit from getting them out of the way. One was geology; the other, archaeology.

A major early geological lidar campaign came about by accident in the mid-1990s. The city of Bainbridge, on Bainbridge Island across Puget Sound from Seattle, asked the Kitsap Public Utility District for an elevation map to better understand creeks and drainage, groundwater recharge rates, and other things. The request landed on the desk of Greg Berghoff, a cartographer with Kitsap. Creating a map based on standard surveying equipment on tripods would take years. But a recent visit by a team from a startup in nearby Belfair, Washington, Airborne Laser Mapping, was still fresh in his mind. The CEO, David Ward, had left behind a couple of floppy disks with digital elevation maps (DEMs) of a gravel pit they'd scanned.[34]

The disks clicked and whirred and finally fed images to his screen. Berghoff considered the ghostly contours. This could actually work, he thought.

Airborne Laser Mapping had bought an early Optech terrain-mapping lidar. Starting in the late 1980s, Allan Carswell's team had worked with two University of Stuttgart professors on a prototype airborne terrain mapper, basing the design on a lidar Optech had built for the Canadian Space Agency a few years earlier. The work involved integrating GPS positioning as Krabill and Martin had done, as well as developing filtering software to identify objects and digitally remove vegetation. The professors, Peter Friess and Joachim Lindenberger, started a company, TopScan, in 1992. Optech developed its first commercial airborne laser scanner to their specs, basing the design on a lidar Optech had built for the Canadian Space Agency a few years earlier. In 1993, the TopScan ALTM, capable of two thousand pulses per second, went into service over Germany. Optech licensed TopScan's positioning and filtering software; shrank the prototype into a lighter, smaller package; and took it to market as the ALTM 1020. Airborne Laser Mapping bought the first one.[35]

Airborne Laser Mapping flew over Bainbridge Island with the Optech lidar in December 1996. Berghoff got the data a couple of months later and put them all into a program that, with his help, generated a map. It was a gorgeous thing: land close to sea level in lime green, graduating to yellows, oranges, and reds toward the top of the island's hills. The TopScan-developed software had shaved away the vegetation and structures. One could see how Pleistocene glaciers had scraped the island's features into a north-south pattern. But down toward the bottom of the image, just south of Blakely Harbor, Berghoff spotted something strange. It was as if a blade had sliced crossways through a bunch of hills. "Why is this going sideways and everything else is going north-south?" he wondered. Was it a road cut? But he knew there were no roads there. Then he noticed the offset drainages—creeks whose meandering paths sharply detoured, then continued.[36]

Bainbridge Island is part of the Puget Lowland to which Seattle and Tacoma also belong. Geologically, it's similar to the Kobe region of Japan, where a 1995 earthquake killed 5,500 people and caused $150 billion in damage. A Puget Lowland earthquake 1,100 years ago caused uplift as high as twenty-three feet, a tsunami, and thousands of landslides and avalanches. But no surface faults had ever been found. Knowing where faults are helps predict how future earthquakes might play out.[37]

Berghoff printed a big version of the map on a plotter. The office was abuzz about this cool thing he'd created, but also about the apparent fault scarp. Word got out. Craig Weaver, a geophysicist with the US Geological Survey and the University of Washington, called Berghoff and asked to check out the lidar-based map. Weaver had used a sonar vessel in Puget Sound to locate an earthquake fault dated to about 900 CE. He had traced it right up to the Bainbridge Island beach. But Weaver had not been able to find it on land. On Berghoff's map, he recognized it immediately: the Seattle Fault.

Weaver reached out to David Harding at NASA, who had flown lidar over the Pacific Northwest with Lefsky and others. Despite all his work with trees, the geologist remained interested in lidar's use in his nominal field and saw opportunity. He and Weaver marshaled federal support of what would become the Puget Sound Lidar Consortium, which within a couple of years had mapped a region 2,350 kilometers square (580,000 acres) surrounding the Seattle Fault zone on either side of Puget Sound.[38]

The Kitsap PUD is still using lidar data—the region has been re-flown since, and Kitsap County and the Olympic Peninsula are scheduled to be scanned again in 2018, Berghoff says. It's no more about earthquakes than that original 1996 survey was.

Lidar helps the utility understand the pressures on either end of a pipe from a reservoir, determine the paths of fiber-optic lines, and establish the locations and sight lines of cell phone towers. "We use it almost daily for our engineering," he says.

A water utility cartographer's curiosity about a new technology led to a deeper understanding of one of America's most worrisome earthquake zones. It's only fitting, then, that a disaster recovery mission led to the first lidar-based map in archaeology. In March 2000, a University of Texas at Austin team led by James Gibeaut flew UT's lidar over the coast of Honduras as part of a Hurricane Mitch impact-assessment mission. UT had been scanning the Texas coast with an Optech ALTM since 1997 to estimate beach erosion and sand loss and also understand hurricane-related flooding.[39]

When their schedule allowed, the UT team took side trips to scan the Copan archaeological site near the Guatemalan border. They stripped out the vegetation using Optech's and their own algorithms, creating the first digital elevation model of an archaeological site. It matched up nicely with a ground survey conducted earlier by Harvard University researchers.[40] Similar work started happening elsewhere. In 2001, UK researchers used an Optech lidar—now capable of scanning terrain with thirty-three thousand laser pulses per second—to survey the region surrounding the Stonehenge World Heritage Site. They digitally removed the foliage to spot extensions to known field systems and barrow cemeteries as well as entirely new sites. The authors concluded that they believed lidar would, in archaeology, "be as significant as the introduction of aerial photography was in the 1920s."[41]

The turning point for archaeological lidar happened back in Central America. John Weishampel, the NASA Goddard postdoc whose dot-connecting had led to Michael Lefsky's pioneering work on lidar in understanding forests, had moved on to the University of Central Florida by 1998, when he worked with a NASA Goddard mission to fly the LVIS instrument over the La Selva Biological Research Station in Costa Rica. This was in preparation for the Vegetation Canopy Lidar NASA mission. The idea was to see how the space lidar might deal with a range of land cover, from grasslands to dense tropical forest.[42] The instrument performed well, bringing back ground returns such as unknown streambeds through the thick forest cover. Weishampel continued working on lidar missions related to forest canopy structure. He also got to know University of Central Florida archaeologists Arlen and Diane Chase.

The Chases had been doing field research at the Maya Caracol site in Belize for twenty years. From January through March, they and a team of about thirty people had been bushwhacking and digging through the rain forest, slowly uncovering more and more of what seemed to them to be a large Maya city. Other archaeologists were skeptical. The terrain was too hilly for much agriculture. Where would they have grown enough food? The Chases' response: they must have terraced it. Indeed, they found thousands of remnants of agricultural terraces in addition to causeways and ruins of residential areas. And while Landsat images hinted that the settlement may have extended six miles or farther out of the center of Caracol, they couldn't find the edge of the settlement when they looked with boots on the ground.[43] At one point, Arlen Chase showed the Landsat images to Weishampel.

"Not to make it sound too bad, but it was pitiful," Weishampel recalls. "Arlen said, 'From these images, we can see these roadways.' And I was like, 'Bullshit. You can't see anything.'"[44]

Weishampel looked into the literature on lidar in archaeology. There wasn't much: the University of Texas group's work in Honduras (they had tucked away their Copan sidelight toward the end of a paper devoted to the primary mission of hurricane damage assessment), the British scans at Stonehenge and elsewhere, some German flights over ruins from the Middle Ages. Lidar could spot archaeological ruins though the canopy in these places. It should work at Caracol too. Starting in 2005, Weishampel and the Chases applied for grants to pay for such flights.

It was slow going.

"We went through like five iterations because everyone told us we were crazy," Arlen Chase says.[45]

The Chases were not the sort who give up easily. These were people who for years had camped in the jungle for months at a time, shooing snakes and bugs and whacking through underbrush to create sight lines for surveying equipment they lugged up and down so many hills. They had seen enough to argue that Mesoamerican cities were as grand as those anywhere else in their era and that Caracol might, at its peak in about 650 CE, have housed 115,000 people, maybe more. They had recently spent three years trying to uncover the extent of agricultural terracing necessary to sustain so many mouths.

"My god, it was all we could to do get two square kilometers of terraces mapped," Chase says. It wasn't enough to quiet the skeptics.[46]

Finally, in late 2007, NASA came through with enough grant money to enlist the National Science Foundation's National Center for Airborne Laser Mapping, or NCALM. NCALM couldn't fly the mission during the 2008 dry season, though, because it was upgrading its Optech lidar to the latest version, one capable of scanning at a hundred thousand pulses per second. A Cessna carrying the new hardware took to the skies over Belize in late April 2009, covering two hundred square kilometers (seventy-seven square miles) surrounding the center of Caracol—ten times as much as the Chases had hand-mapped over the years. The Optech ran for about nine hours across several flights, taking more than four billion measurements.[47]

Weishampel was in Nebraska correcting Advanced Placement Environmental Science exams when NCALM sent him the Caracol lidar data. He opened the files on his laptop computer. It didn't take an archaeologist to see that the lidar had done its job. "You zoom in and it's unbelievable," he says. "It's an epiphany." On his laptop screen he was seeing Caracol as no one had in more than a thousand years. Back in a University of Central Florida conference room, Weishampel projected an image derived from the lidar data onto a pull-down screen. It was quiet for a moment. Arlen Chase then spoke.

"Holy shit," he said.

Weishampel helped Chase navigate the geographic information system (GIS) software to peruse "this world that he had spent decades in but couldn't see." Chase saw features he had discovered years ago and ones a few feet from them that he had missed. There were causeways and the remnants of dwellings. And there were thousands and thousands of terraces—90 percent of the landscape had been modified by people who used no wheeled transportation and kept no beasts of burden. There were clusters of *plazuela* dwellings, reservoirs, causeways, caves, and much more that were new to the couple who knew the place like no one else.[48]

**A lidar scan of Caracol's Puchituk Terminus taken in 2009. The extent of the site and its agricultural terracing helped change archaeologists' views on the size and sophistication of Maya civilization. (Image courtesy of Arlen and Diane Chase, Caracol Archaeological Project.[49])**

"Just a few days of flyovers and three weeks of processing yielded a far superior picture of Caracol than on-the-ground mapping ever had," the

Chases and Weishampel wrote. The implications extended beyond Caracol. The Caracol lidar data changed fundamental assumptions about Mesoamerican civilization by proving that the city was big, sustainable, and interconnected. "For too long, Maya archaeologists have been blinded by the jungle, able only to sample once-wondrous cities and speculate about vanished people," they continued. "The airborne LiDAR data will help us finally dispel preconceived notions about ancient tropical civilizations—that they were limited in size and sophistication—by letting us peer through the trees."[50]

Others caught on quickly. Weishampel presented results at a remote sensing meeting in India in late 2009; University of Sydney archaeologist Damian Evans was among the thirty or so people there. At the same time, the Chases were showing the images to archaeological audiences, including Evans's advisor, Roland Fletcher, who led the Greater Angkor Project in Cambodia. Fletcher and Evans soon had a grant to fly a helicopter-mounted lidar system over Angkor, which they did in early 2012. Others have followed in Southeast Asia, Mexico and Central America, Europe, India, the Middle East, the United States, and elsewhere. Also in 2012, documentary filmmaker Steve Elkins, inspired by the Caracol lidar work, enlisted NCALM in a search for the legendary lost "White City" in Honduras's Mosquitia rain forest. They found an expanse of ruins.[51]

In 2013, the Chases and other archaeologists landed an Alphawood Foundation grant to have NCALM scan another 1,000 square kilometers (386 square miles) of western Belize near Caracol. In 2016, the Guatemalan Pacunam Foundation kicked off a three-year lidar campaign in which NCALM would fly 14,000 square kilometers (5,405 square miles) of the same Maya lowlands to which Caracol belongs. The first year they scanned 2,100 square kilometers (810 square miles), uncovering sixty thousand ruins, including pyramids, causeways, quarries, dwellings, tombs, and defensive fortifications, further suggesting that Maya civilization in Central America was "more comparable to sophisticated cultures such as ancient Greece or China than to the scattered and sparsely populated city states that ground-based research had long suggested."[52] It was among hundreds of lidar archaeology missions that have flown since that first flight over Caracol in 2009.

There's still good old-fashioned fieldwork to do on the ground in Caracol and elsewhere to investigate the secrets lidar scans have revealed. Lidar's proliferation has brought questions as to whether detailed lidar-derived archaeological maps should be publicly available—and thus available to looters. Regardless, there's no doubt that lidar is safe in its position as an indispensable tool in understanding forests; their implications on biodiversity, climate change, and power infrastructure; and the treasures long hidden beneath them.

# CHAPTER 10
# GLASS HALF FULL

Lawrence Crowder gives me directions over the phone. He starts with straightforward lefts and rights along two-lane dirt roads a few miles southwest of Alamosa in Colorado's San Luis Valley. Look for a central-pivot irrigator, he tells me, where I'll find a double-track path, which I should take. Then look for two red tractors out raking alfalfa. He'll be in one of them.

On my second pass, I find the path and spot a couple of tractors way off beyond it. I park well short of the hub of the irrigator. A red-tailed hawk perched on it waits for a mouse. The closer of the tractors follows brown and green stripes that expand in huge concentric circles defined by the path of the irrigator without which the alfalfa would stand no chance. I walk toward the field's rim, stepping high through bands of shorn alfalfa and its stubble.

As mine and the tractor's paths meet, the tractor stops. A man with short-cropped hair and a mostly gray beard about the same length dismounts into the midmorning August sun. He wears jeans and a blue T-shirt. Crowder shakes my hand and invites me into his tractor. His dog, a papillon named Kadee, is curled into a denim shirt near the foot pedals. He starts the engine again, and it gets loud. We holler over the racket as the tractor bounces through the field. Behind us, eight medievally spiked wheels whirl, pushing fat strips of felled alfalfa into a single, deeper column like a boat wake in reverse. Tomorrow, a baler will inhale it and cough up four-by-four-by-eight-foot blocks weighing close to a ton. The bales will go to dairy farmers in New Mexico and Texas, where cows will turn them into milk. Alfalfa comprises the bulk of Crowder's 600 acres. His other big crop is barley, about 150 acres of it. Coors turns that into beer, Crowder tells me—has for about forty-five years, since Crowder's dad landed the contract in the early 1970s.

**Lawrence Crowder and his faithful sidekick, Kadee.**

There's some irony in the fact that crops from a valley that averages a little over seven inches of rain a year (about one-sixth of the US average) ultimately yield the liquids of school lunches and happy hours. But so it is in the San Luis Valley, and this valley isn't the only place relying on water from somewhere wetter to produce what we eat and drink. About 1.2 billion people—one-sixth of the world's population—depend on seasonal mountain snowpack. Nearly three-quarters of the water in the western United States trickles from these frozen blankets. Even if you're reading this in a rainier part of the country, it affects you: more than one-third of US vegetables and two-thirds of our fruit and nuts grow on snowpack-irrigated California farmland.[1]

Even among farmers in what would otherwise be greasewood-studded desert, sixty-six-year-old Crowder is in a particularly good position to talk about irrigation. He's been the president of the CommonWealth Irrigation Company since 2001. The company's two hundred shareholders hold the water rights to Empire Canal, the second-largest irrigation ditch coming off the Rio Grande. "Water rights" gets us into water law, which makes laser physics seem easy. But it boils down to the Rio Grande being vastly oversubscribed, in Colorado and on down through New Mexico and Texas.

In 1938, right around when Crowder's grandfather homesteaded here after the Dust Bowl chased him out of eastern Colorado, those three western states signed the Rio Grande Compact, which guarantees certain flows at the New Mexico and Texas borders. The flows depend mostly on how much water is expected during the irrigation season, which runs from April through October. That flow, in turn, depends on how much snow has accumulated in the San Juan Mountains to the west that fill the Rio Grande. The snows there are the reason it's so dry here. The snow adds up to an average of maybe fifty inches of precipitation a year across hundreds of square miles. It melts into a lot of water. But just how much has been anybody's guess.

Which is a problem for someone like Crowder. Water rights are seniority based. The older the water right, the closer you are to the front of the line to irrigate. The oldest of CommonWealth's five water rights dates to 1882, the newest to 1890. Old, you'd think. Turns out those are "basically junior rights," Crowder says, "but we're high enough on the junior list that if we're in good water, we run pretty good priority." The oldest water right on the Rio Grande was appropriated in 1866, ten years before Colorado became a state. The state's spreadsheet tracking Rio Grande water rights has 640 rows, the most recent entry being from 2007.

Depending on the year, Colorado delivers between about 25 percent and 70 percent of the Rio Grande's water to New Mexico (the less snow, the lower the percentage). Last winter was an average snow year, so Colorado will be sending 29 percent of the river on to New Mexico in 2017, about two hundred thousand acre feet.[2] The other five hundred thousand or so acre feet were allocated based on the seniority of the water right, starting with that 1866 right and working its way to Crowder's most senior one, which is 104th in line, and on down. If there's a lot of water, more junior rights see their ditches fill for thirsty fields. If not, the gates close when state officials estimate they have to close to deliver two hundred thousand or so acre feet to Colorado's neighbor to the south by the end of the year. Everybody behind that spot in line gets nothing.

If one knew how much snow had settled across a vast mountainous region, and how much water was locked up in that snow (called snow-water equivalent), and how that snow had blown and piled and drifted into sunny and shady spots, and how the weather and dust and the sun would melt that snow into runoff, this all would be easy. Farmers across the San Luis Valley's six hundred thousand irrigated acres could plan on having water or having no water as the growing season drew on. If you knew you had no water coming, you wouldn't plan on a third cut of alfalfa, say—it's a deep-rooted plant that can survive through dry times, though it won't grow enough to produce decent hay. Or, if you don't have well-water backup, as Crowder and many others do, and you farm grains, potatoes, carrots, or lettuce, you maybe don't plant at all.

But none of these snow-related variables are known to any exactitude. In fact, projected and actual flows have typically varied by 30 or 40 percent. So they'll send too little downstream in April and May when flows are strong and have to make it up in June and, in the increasingly rare years the Rio Grande can sustain it, in July. So the line gets cut off earlier than expected in the hot, dry months when the plants are thirstiest. The irony is, the more senior rights holders, who with better water forecasts would have gotten more water, suffer the most.

"The last couple of years, I feel like our canal's gotten hammered when it shouldn't," Crowder says.

Not all of his land has well-water coverage, he says. "If I was wanting to plow that out and the forecast comes out and says, yeah, we're gonna have a good year, and I put in alfalfa seed and then all of a sudden it drops off and I can't irrigate it, then I've just wasted my time and money on that."

He's quiet for a moment, and I wonder if a NASCAR driver could follow these lines any better than Crowder in his tractor. Then I mention a research project happening right now, one involving NASA, using a high-end lidar and another instrument. It could change the water management game.

"If there was a better way of doing that, you know, to me it'd be well worth it," Crowder says. "And we've got a lot of shareholders who don't have any wells. And so they're at the mercy of the Rio Grande River. And Mother Nature. And Craig," he adds, laughing at this last bit.

Craig Cotten is the division engineer for the Colorado Division of Water Resources district in charge of the Rio Grande basin. He works from a low-slung office building in Alamosa, a town of about ten thousand in the south central part of the valley. In a frame high on the lobby wall is a quote from 1890 attributed to J. P. Maxwell, the state's top water engineer at the time: "He who expects the letter of the law in relation to irrigation to be executed with the precision of clockwork, and that infallible results will be obtained, has a small conception of the tangled web of difficulties in the way, and meager knowledge of the uncertainties of the element to be manipulated."

Cotten is a big guy with a handlebar mustache and a pleasant demeanor, the latter of these at least as important in this job as his technical and managerial skills. He and the eleven water commissioners working for him are like kindergarten teachers with only enough snacks for maybe half of the kids in their classrooms, if that. Except the kids in this classroom are lawyered-up farmer-entrepreneurs whose livelihoods depend on the snacks.

"The biggest problem we have is we don't have enough water to go around," Cotten says. "Every day of the irrigation season, there's a ditch owner that's not getting their water because we don't have it somewhere in the valley. This time of year, three-quarters of the ditch water owners are not getting their water."

The issue boils down to a contractual obligation clashing with the mysteries of nature. The interstate compact establishes the Rio Grande flow that must go to New Mexico by December 31. But how generous the San Juan Mountains will be has been hard to pin down. That's not for lack of trying. The status quo, Cotten says, is to look at the snow-water equivalent numbers across the dozen or so SNOTEL sites—snow telemetry stations—up in the San Juans, compare them to some historical year, and base the flow estimates on that. "This year, it's an average of twenty inches of snow-water equivalent. Back in 1972 we had twenty inches, and this was the runoff we had," Cotten explains. "It's a real simple analysis, and it's been the standard for a lot of years."

Complicating matters is that the few reservoirs on the upper Rio Grande are privately owned, so Cotten can't fill storage in fat times and release it when things dry out come July and August. So forecasting is even more important here than, say, on the upper Colorado River, where there are massive reservoirs. NOAA computer models attempting to help forecast flows have been hit-or-miss, Cotten says.

And so, as Crowder tells me, "Craig's hands are tied because he's trying to guess how to set the curtailment and how not to put too much over down on the state line. He's going by the best guess he's got. And that's basically to me what we got: we're guessing. And we thought earlier this year that we were going to run to the tenth or fifteenth of July. And that river just dropped like a rock."

This is where the NASA program with airborne lidar and another instrument—a spectrometer that sees in seventy-two colors, from the visible into the near infrared—comes in. It comes in thanks to Joe Busto.

Busto is a scientist with Colorado's Water Conservation Board. Whereas Cotten and colleagues are on-the-ground administrators and rule enforcers, Busto's mandate is to make sure Colorado conserves and protects the water it has and plans for the future. The problem in the Rio Grande basin, Busto saw, wasn't NOAA's models. It was that there were all of a dozen SNOTELs up in the San Juans, and they were between 9,500 feet and 11,600 feet in elevation despite huge amounts of snow up higher, toward peaks in the range of 13,000–14,000 feet. There were almost no data to put in a model.

"It's ridiculous to miss by 20, 30, 40 percent of volume every year. That's a sixty or a seventy on your exam. That's a C-minus or a D," Busto says. "People have paid for that water. When the numbers aren't right, it's a big deal to them. If we had gotten the numbers better, they would have gotten their water. If we can all work together and implement better methods, we need to do it, and we need to do it now."[3]

He brought in a mobile weather radar to quantify snow as it falls. It's helped some. He also came across this NASA program with the lidar and the spectrometer. It's called the Airborne Snow Observatory, and it was Tom Painter's idea.

Painter is a NASA JPL scientist with mountain snow in his marrow. He grew up in Fort Collins, where his dad was a Colorado State University math professor who coached the CSU ski team on the side. When Painter was six, his dad took the family to the Matterhorn Museum in Switzerland, where he met the granddaughter of Edward Whymper, the first to summit the iconic peak. "That just cemented my love of the mountains," Painter says. He finished his math degree at CSU and spent a season skiing in Utah, where he came to realize there were careers in studying snow. His PhD focused on using spectroscopy to understand what's happening with mountain snow. Photos of the Matterhorn and Long's Peak in Rocky Mountain National Park, visible from Fort Collins, graced the cover of his University of California, Santa Barbara dissertation.[4]

Painter's scientific specialty was in linear algebra algorithms to very closely interrogate the reflectance of a snowpack. Melting and freezing hits the tiniest snowflake crystals first. They form grains of varying sizes, which reflect light differently, which a spectrometer can pick up. Snow grain size is tied to snow-water equivalent—the larger the grains, the more concentrated water in the snow. Spectrometers also pick up reflectance from dust and soot from snow-packs. That affects how much sunlight gets absorbed rather than reflected and drives the pace of snowpack melt, Painter says. Higher temperatures driven by global warming get the scientific attention, he says, but more than 90 percent of energy that goes into melting snow comes from absorbed sunlight.

"It's a much bigger signal than the warming," Painter says. The snowpacks he's studied are now five to seven times dustier than they were before the mid-1800s, mainly from land disturbance to the desert system. It's the same story with "the plains of Patagonia, the deserts of Central Asia and the Indian subcontinent, the deserts next to the Andes, the Taklamakan Desert, and the Gobi Desert," he adds.[5]

Painter's work in spectrometry showed the tool could provide insights into how much water a given bit of snowpack holds and how the sun might melt it over time. The missing piece was how much snow there was. For that, you need to know the snow depth. Radar did a fair job with shallow snow over flat terrain, but it was of little use over the mountains, where it matters most.

As with lidar in archaeology, the answer came from forests. Painter attended a talk by Greg Asner, a Carnegie Institution of Science researcher. Asner, a tropical forest ecologist, had led the creation of the Carnegie Airborne Observatory, which had been flying a lidar-spectrometer combination over tropical forests to assess biomass and forest composition, which a spectrometer can pick up because different species' leaves reflect light differently.[6]

As Painter listened, he realized the same approach could work for snow. The density of a snowpack doesn't change much across the landscape. The big variable is snow depth. If you were to fly a lidar before the first snowfall to capture the landscape itself, then again in the winter, you could subtract "snow on" from "snow off" and have your snow depth across a landscape. Combine that with spectrometer data about snow grain size and dust, and you might gain unprecedented insight into how much water was socked away and how fast the sun might melt it.

"Thank god for Greg Asner and his love of tropical ecosystems, which has allowed this to happen," Painter says.

Painter needed a lidar guy. He knew one: Jeff Deems, of the University of Colorado–NOAA Cooperative Institute for Research in Environmental Sciences (CIRES). Deems was a reformed ski bum like Painter and, like Painter, also a snow/watershed science specialist. Painter had been on Deems's graduate committee, and Deems's PhD thesis involved working with lidar data collected over Colorado in 2003.[7] The two were good friends now. With Painter leading the mission and focusing on the spectrometer and Deems spearheading the lidar, they built the Airborne Snow Observatory (ASO) and started flying it in 2012. California's Sierra Nevada range has been the focus. But Joe Busto worked with his own bosses and Painter to bring the ASO to Colorado. Two months before I rode in Lawrence Crowder's tractor, the ASO flew over some of the snow that helped grow the alfalfa he was raking.

Shortly before the Beech King Air 90 is to take off on this second Saturday in June, it is eighty-five degrees on the tarmac, headed for ninety-five. It's the sort of heat that melts the very notion of snow into an abstraction. Indeed, here in Grand Junction, Colorado, the snow is long gone. The ASO flight team chose it because the hotels in Pagosa Springs, their preferred launch point for scanning the snows of the Rio Grande headwaters, were booked solid for a cycling event. So they will commute for an hour to Colorado's deep south, scan for about three and a half hours, and commute back. The team flew in from the ASO's home base in Mammoth Lakes, California, yesterday and today are planning on spending about ten hours in the air on two flights over snowy peaks. Between the flights, they'll take a quick lunch break back in the heat of western Colorado.

Painter is in California. Deems is hiking around high in the San Juans, digging holes in the snow and taking successive slices from the surface to the ground. This is to measure actual snow-water equivalent for a reality check on what the technology in back of the King Air 90 will estimate. He brought his skis along, but given how late it is in the season, he won't need them. It's warm up there too—in the sixties—and sunny.

The NASA ASO flight team consists of four people. Two are pilots Josiah Grindrod and Sam Wilson, who work for an air services company, Dynamic Aviation. It owns the plane, which is white with red and blue stripes and oil streaks like crow's-feet, to be expected from something built in 1971 and flown hard. NASA JPL science technologists Elizabeth Carey and Peter Lawson run the scientific hardware from two seats just behind the pilots. That hardware, which is worth several times more than the aircraft carrying it, occupies much of the back of the small plane.

Carey, whose purple hair highlights match her eyeglass frames and her Fitbit, walks me up the steps to the cabin to show me the instruments. She joined JPL with a master's in physics and, when she's not flying with the ASO, does things like making simulated comets in a vacuum chamber. The lidar, a Riegl, scans with two laser beams, each of which sends out four hundred thousand pulses per second and will measure the snowpack to an accuracy of less than four inches. It also has a camera. Mounted just behind it on the same inch-thick black metal plate is the spectrometer, and next to that is an electronics rack with supporting hardware for these boxes.

They'll fly at twenty-three thousand feet above sea level despite the cabin not being pressurized, so they'll wear oxygen masks fed through three beat-up green oxygen tanks strapped between the seats. Carey has brought a thick winter coat despite the heat because it's cold up there. And there's no bathroom.

This was a bigger hurdle than anything technical when he was pitching the idea of ASO, Painter says. JPL has flown instruments for years. Painter was thinking five hours at a time from the start—you needed that to fly enough lines to cover enough of a basin to deliver useful data. "They said, 'No, no, no, your team is not going to be able to deal with five-hour flights. It's too difficult on the body. The most you're going to be able to fly is one and a half to two hours,'" Painter says. "Sure enough, they fly five-, five-and-a-half-hour sorties at twenty-three thousand feet on oxygen, and they'll do it multiple times a day. I think one part of it was rethinking what people thought was possible. The other is having the right people who were willing to just take on that challenge and go do it."

**Elizabeth Carey of NASA JPL's Airborne Snow Observatory team points out the Riegl lidar from below the King Air's belly in Grand Junction, Colorado.**

He found the right people in Carey and Lawson. Yes, she has to lift her oxygen mask to nibble on Goldfish crackers, and it's freezing, and everybody beelines for the restrooms when they land. But it's more than suffering for the cause, it seems. "It's a blast," Carey says. "I mean, you're ten thousand feet over a mountaintop." Lawson, an optics PhD, says pretty much the same thing: "Flying's fantastic. And we're flying over the most beautiful parts of the United States." Grindrod, the pilot, says this is harder work than, say, flying rabies-baiting missions over the Appalachians or gravity-measurement missions over Alaska, where autopilot does much of the flying. Here, he and Wilson alternate at the controls every hour to stay fresh. But he, too, describes ASO flights as "fun." Plus, he adds, "We're doing something that's worthwhile."

When they arrive at the basin, they fly a roughly twenty-mile-long line at about 230 miles per hour, U-turn, and head back in the opposite direction on a parallel track, covering perhaps five hundred square miles in a generally north-south lawn mower pattern dictated by the spectrometer's preferred sun angle (it needs sunlight to reflect off the snow; the lidar would be just as happy in the dark). The flight yields about five hundred gigabytes of data, she says. The ASO team will, thanks to software they developed, combine the lidar's and the spectrometer's outputs into an "intermediate data product," as Painter calls it, within twenty-four hours. That's really fast given the complexity and volume of data involved, and it took months of software development to get there, he says. The intermediate product feeds into a model that produces the streamflow forecasts that water managers like Craig Cotten can use to decide whether farmers like Lawrence Crowder get their water or not.[8]

Water managers in California, where ASO flies over the Tuolumne River basin and others on a weekly basis from late winter into early summer, are already benefiting, Painter says. The Tuolumne snowpack feeds the Hetch Hetchy system, among others, which supplies water to three million people in twenty-nine cities across the San Francisco Bay area. With ASO having flown for five years running now, he says, data from April flights can help predict July runoff within a percent or two of actual flows. So water managers know, for example, how much they can release to generate hydroelectric power earlier in the summer without worrying about water supply in San Francisco months later.[9]

The NASA JPL Airborne Snow Observatory team of about twenty-five scientists, remote sensing specialists, instrument operators, project managers, hydroelectric modelers, and computer experts is expanding to cover into new basins in California and elsewhere. It's still a research system, and for now it's all on that one King Air. Like the Airborne Oceanographic Lidar before it, the ASO is more about establishing what's possible than expanding operations to mountain basins around the world that desperately need better streamflow predictions. Painter believes commercial systems will emerge to fulfill those needs, eventually, just as the Riegl lidar his Airborne Snow Observatory depends on—and ones like it made by Optech, Leica, and others—followed, if indirectly, from the pioneering work of the AOL team at NASA Wallops.

The Rio Grande flights have been less frequent than those in California. But Joe Busto, the Colorado Water Conservation Board scientist, says he likes what he's seen. "The part where I went *aha* was, we never knew what the damn snowpack was—we just knew there was an area that was white or wasn't," he says. "Now we've got gory gobs of spatial detail, in watersheds we didn't have before in wild areas we'd never explored."

Craig Cotten says the National Weather Model that the ASO data feed into has been more accurate than its predecessor, though it's still early days. "It really seems like it could be something we could grab ahold of and use to improve our forecasting," he says.

Back in the alfalfa fields southwest of Cotten's office, Lawrence Crowder stops his tractor to let me off. It's time to switch to the baler, which he'll run until it gets dark. Kadee the papillon will keep him company. I step back as the rake folds in like the wings of a landing bird, and Crowder rides off to the west, toward the mountains that feed the river that feeds fermenters in the bellies of cows and in the bowels of a giant brewery in Golden. Tom Painter's ASO can't make it snow. But it just might plow through the uncertainties that make predicting the Rio Grande's flow so very difficult for Cotten and colleagues. And that would make it easier for the likes of Lawrence Crowder to make a living as they work to put food (and drink) on our tables.

# CHAPTER 11
# FIREPOND

Tom Painter's Airborne Snow Observatory is another example of NASA's pivotal role in the advance of lidar, one reaching back to the retroreflectors on Explorer Beacon 22 and Milton Huffaker's Apollo-era experiments. Another US government agency has done at least as much for the technology: the Department of Defense.[1] Before there was lidar, DoD poured money and talent into radar, maser, and laser development; it paid for the Luna See telescope, its Raytheon laser, and Fiocco and Smullin's time; it bought systems from Hughes and others; it paid for the Canadian detectors NASA Goddard ended up building into so many space instruments; and it fostered the development of countless other systems that pushed lidar ahead in areas as diverse as wind detection, aerial mapping, and object discrimination. The work continues today, a lot of it under wraps.

The most famous use of military lasers has to do with smart bombs. It's not quite lidar: an aircraft or soldier on the ground aims an infrared laser at a target. A bomb dropped from an aircraft detects the light and homes in. Texas Instruments developed the first one in the mid-1960s, yielding the Paveway system, which debuted in 1968 during the Vietnam War. In 1972, a group of F-4 Phantoms aimed twenty-six laser-guided bombs at the Dragon's Jaw bridge at Thanh Hóa, a key North Vietnamese rail and road supply line that had survived 871 sorties and cost eleven US fighter jets. This time, the "toughest target in North Vietnam" went down.[2] The system revolutionized tactical air-to-ground warfare, sharply reducing the number of bombs needed to destroy a target as well as minimizing collateral damage and the risk of getting shot down. Evolved versions of Paveway featured in the Gulf War and are still used today.[3]

For some sense of how military lidar evolved, I talked with two men who have been involved with it since the 1980s, Gary Kamerman and Paul McManamon. Kamerman comes from the defense contractor world; McManamon is a longtime Air Force Research Laboratory optics researcher and leader. Both have security clearances and know more than they can talk about. Neither has anything close to a comprehensive sense of the US military's various lidar-related activities. That's because nobody does. Kamerman recounts a conversation he had with McManamon.

"I said, 'Paul, you've seen an awful lot that's happening here, but I can tell you for a fact that there are laser radar programs you're not briefed into.' And he looked at me and said there were laser radar programs that I wasn't briefed into," Kamerman says. "We realized that there was no one person in the entire country that really knew what was going on."[4]

Kamerman *can* talk about a couple of interesting early military lidar projects having to do with cruise missiles. The first was a research effort at Eglin Air Force Base in Florida. During the Cold War, military planners expected a Soviet invasion of Western Europe to motor in from East Germany via the Fulda Gap. Such a force would include thousands of tanks and other military vehicles. There would be far too many of them to destroy without resorting to nuclear weapons, so the focus would be on the tanks. Huge numbers of A-10 Warthog tank-hunting aircraft would be lost. So the idea at Eglin was to put a 3D imaging lidar in the nose of a cruise missile. The lidar would compare topography and landmarks with stored maps and avoid obstacles en route, then spot tanks and lob forty downward-firing mortars at them.[5]

This all would be ambitious in today's Xbox world, much less in the Atari era. It didn't go very far. But another cruise missile program, called the Autonomous Terminal Homing system (ATH), ended up in an actual military system—at least, part of it did. The Air Force Research Laboratory program, launched in 1979, was based at Wright-Patterson Air Force Base in Dayton, Ohio, and involved several military contractors. The idea was to use lidar to improve cruise missile guidance, a problem scientists and engineers had struggled with since the "flying bombs" of the World War I era.[6]

Cruise missiles fly far and hug the ground to avoid air defense radar. The challenge is to keep them from crashing and then to bomb just the right place, two thousand or more miles away. While designed with tactical advantage in mind, the prospect of superaccurate cruise missiles had huge strategic implications too, Kamerman says. The need to target Soviet missile silos with nuclear weapons was premised on the fact that with limited accuracy, you needed a nuclear warhead to ream out a two-hundred-foot-deep crater wide enough to ruin the opposing silo. But if you could hit the silo's door, a thousand-pound conventional warhead would do the trick, he says. You could, then, deter a nuclear opponent with nonnuclear weapons.

The ATH program became a precision targeting competition between a $CO_2$ lidar and an infrared camera, both of which could work at night. Albert V. Jelalian's group at Raytheon, a central player in the development of $CO_2$ lidar for tactical as well as atmospheric uses, built the ATH lidar. They both worked well, but the lidar offered advantages in weather penetration, obstacle avoidance, and ease of mission planning. For the follow-on Cruise Missile Advanced Guidance system (CMAG), they went with lidar.[7]

CMAG started in 1983, with McDonnell Douglas and General Dynamics each given contracts to build independent, lidar-based systems. These systems boosted laser power for longer-range imaging, processed 3D data to avoid obstacles and identify targets, and kept precise track of speed based on the lidar Doppler effect (being coherent $CO_2$ lidars, they could harness the same physics that Huffaker's wind-sensing lidar did, if in a different context). Both teams were successful, says Ronald Kaehr, the air force's CMAG project manager. But by 1987, it was clear that CMAG would be pricey and that alternatives based on then-new GPS would be cheaper and easier to employ. The program was canceled.[8] A bit of the lidar survived, though, in the stealth, nuclear-tipped AGM-129 Advanced Cruise Missile built by the same contractors. It relied on GPS and a radar, but the navigation system did include a lidar-based Doppler velocimeter to keep precise tabs on how fast it was moving.[9]

McManamon's most influential work in lidar started around the time CMAG wrapped up. With the cruise missile programs, the idea was to use a nascent technology to improve an existing weapon. McManamon was interested in inventing a technology that could revolutionize all sorts of different platforms. Like the laser itself, the technology McManamon wanted to create—something called an optical phased array—had its roots in radar.

The classic radar unit rides a gimbal, its single transmitter sending out microwaves and tuning in for a response with its big synthetic ear. A phased array radar spreads its coverage around without physically moving at all. It consists of a line (or a sheet, depending) of transmitters that can steer the direction of their collective output by slightly delaying the timing of their individual pulses.

The effect is much the same as slapping a floppy foam swim noodle onto the surface of a swimming pool. The noodle closer to the hands will generate its part of the wave slightly before the end of the noodle. The result propagates off to the sides, but also slightly forward of the noodle-slapper. Phased arrays can work in both directions (or many directions, if arranged in a plane).[10] They're also great as weather and defense radars, on ships, on combat aircraft, and on missiles. They're hard to make, but hard to break too, and can be formed into funky shapes that hug the surface of a fighter plane or a rocket. They can also shift their beams around much faster than a mechanically steered radar and multitask to track things in different places at the same time.

Without an optical phased array, a laser for mapping or recognizing an object like a tank or doing point-to-point battlefield laser communications or shooting an infrared beam to confuse an incoming heat-seeking missile—they all need moving parts and a good deal of supporting hardware to steer mirrors fast and accurately.

"The optical guys, we always had these mechanically complex steering mirrors. The gimbals, you know, whatever else, and they'll always be complex and they'll always be expensive," McManamon says. "So starting in about '85, I began to say, 'So why can't we do what they do? You know, why can't we steer laser beams with no moving parts?'"[11]

Among the many challenges had to do with size. Each phased array cell needs to be at most half the diameter of the wavelength involved. For a radar with a wavelength like that of your microwave oven, each cell would be about the width of a deck of cards. For an infrared laser with a wavelength eighty thousand times shorter, the cell would be 750 nanometers across. You could line up a hundred of them across the width of a human hair. And into that you'd need to pack electronics, and connect them with wires, and avoid interference and electromagnetic spillage among thousands of neighboring cells.

In 1987, under a contract called Beam Agility Techniques, McManamon brought in Raytheon, Hughes Research Labs, and Westinghouse to see if they could build optical phased arrays. A big focus was on liquid crystal display (LCD) technology—the stuff of laptop computer screens back then and about every screen today. A laser beam would interact with these microscopic LCDs, which would either block or send light through. By the early 1990s,

Raytheon had built what was, in a sense, the world's largest phased array. It was all of an inch a half across, but it had forty thousand tiny elements, more than the ninety-five-foot Cobra Dane phased array ballistic missile tracking radar the company had built for the air force in Alaska some years earlier.[12]

In 1996, McManamon and others, mostly from Raytheon, coauthored what would become the seminal work on optical phased array technology.[13] In 2000, DARPA launched its Steered Agile Beam program, which involved about fifteen contracts with companies and university teams, to build on the work McManamon started. The goal was to cut system size and weight by a factor of thirty. Nearly two decades later, there's still no military system using an optical phased array, McManamon says, though they're getting closer, and he expects it to one day be applied to lidar, laser communications, and other systems.[14] The biggest impact, though, may not be military at all. Solid-state lidar based on McManamon's phased array inspiration more than thirty years ago could be pivotal to the future of self-driving cars.

The story of lidar would be incomplete without talking about the biggest, most powerful lidar there ever was—and probably ever will be. It started with a scientist named Leo Sullivan, and the seed was planted twenty-four years before the project started. Sullivan had graduated from MIT with a physics degree in 1940 and, after a couple of years working on electron microscopes as a research assistant there, hired into the MIT Radiation Laboratory and got involved in the development of the world's first radar-controlled antiaircraft gun system. By 1944, he was in Europe helping install and train Allied soldiers on it. The system worked, even against unmanned German V-1 missiles buzzing toward London—pulse-jet-powered proto-cruise missiles. Against the rocket-powered V-2s, though, its analog computers were too slow. Sullivan held onto the telegram one of his superiors sent him. The boss asked, in essence: "Germans hitting London with V-2 rockets. What are you going to do about it?"[15]

The MIT Rad Lab closed after the war; MIT Lincoln Laboratory emerged in 1952, tasked by the DoD to build a unified radar system to detect Soviet bombers and, later, intercontinental ballistic missiles coming in from over the Arctic. With the advent of the laser, Lincoln Lab scientists recognized the technology's kinship with radar and its potential as an optical version of it capable of thousands of times higher resolution. In 1966, the Office of Naval Research contracted with Lincoln Lab to build a lidar capable of quickly discriminating an incoming nuclear warhead from accompanying decoys from hundreds of miles away. Leo Sullivan led the program, called Firepond.[16]

The system, built at the Millstone Hill radar site in Westford, Massachusetts, involved a continuous-wave (that is, with a consistent, not pulsing beam) laser that was generated at first inside a series of seventy-foot-long glass pipes of the sort otherwise used to move milk around dairies. The laser light bounced off mirrors and up a forty-foot tower to another mirror, then received return signals via the same path. Later, the team added a massive laser amplifier called LRPA to boost power.[17]

Sullivan's team had to invent everything from the detectors to a laser two hundred times more powerful than had ever been built. The project went on for fifteen years, until 1981, when it was canceled because of problems with the LRPA that kept the system from running long enough at high enough power to work like it was supposed to.[18]

That might have been it for lidar at Firepond had it not been for a nationally televised speech President Ronald Reagan gave on March 23, 1983. The focus was on Soviet advantages in nuclear as well as conventional military forces as part of a pitch to ramp up US defense spending. Then, toward the end, the president spoke of harnessing the country's technological might to build a system that "could intercept and destroy ballistic missiles before they reached our own soil or that of our allies."

"I know this is a formidable, technical task, one that may not be accomplished before the end of this century," Reagan continued. "Yet, current technology has attained a level of sophistication where it's reasonable to begin this effort. It will take years, probably decades of effort on many fronts. There will be failures and setbacks, just as there will be successes and breakthroughs. . . . But isn't it worth every investment necessary to free the world from the threat of nuclear war? We know it is."[19]

The multibillion-dollar Strategic Defense Initiative, quickly nicknamed Star Wars, was born and indeed fueled across many fronts—from X-ray lasers to rocket interceptors to be launched from space. McManamon's Beam Agility Techniques contract was among them. Surveillance was a major theme. Lincoln Lab proposed a laser-based system to tackle the problem of quickly discriminating an incoming nuclear warhead from accompanying decoys from hundreds of miles away. This might sound familiar.

Leo Sullivan put off retirement to help out William Keicher, who would be the technical lead this time around. As with its predecessor, the new Firepond would be a research system—a prototype to show that the laws of physics would let it happen. Only then could working units be sent up on airplanes or perhaps even satellites. While some key aspects of the earlier Firepond could be reused, much of it would be ripped out and replaced. One of the big changes would be a new amplifier to replace the LRPAs. Brian Edwards, a thirty-two-year-old engineer, would lead the effort, which started in 1985. He was given just one directive: "This one better work."[20]

Edwards had known he wanted to be a laser engineer from the time he was a kid, not long after the technology was invented. Coming out of his PhD program at the University of Illinois in 1979, he intended to work at Los Alamos, but his wife wanted to go to law school. He took a job at Raytheon west of Boston, where she could do that while he worked on infrared detectors for Sidewinder missiles. He knew little about Lincoln Lab when a college friend working there said they were looking for a laser guy. In 1981, he hired in and went to work on a truck-mounted lidar.

That system, as with the new Firepond, was a Doppler lidar. It imaged things by assembling pixels into pictures, just as a direct-detection lidar of the sort used to map Mars or the Rio Grande Basin would. But it built the picture in Doppler space. If something was moving toward or away from the laser, the system could see it based on the slight shift of the light's frequency as compared to the frequency sent out. A tumbling warhead the size of a small car, and its decoys would be moving fast. A Doppler lidar would, in theory, be able to measure the distances to them as well as image them within fractions of a second as far away as New York is from Chicago, something well beyond what standard telescopes or even direct-detection lidar could hope to do.[21]

Firepond redux would be on the ragged edge of the day's technology, though, and take a team of about seventy Lincoln Lab staff in the newly formed Optical Discrimination Technology Group five years to build. Several times that number worked on the system at contractors. As it came together, it worked like this.

What was probably the most stable laser ever built formed the original pulses.[22] This master oscillator was a holdover from the original Firepond, built into what was known as the Tomb. The Tomb had rubberized walls and gravity-fed water cooling, and its thick granite laser table was mounted on columns driven straight into bedrock, all to minimize vibration. This super-stable beam bounced over to a modulator to imprint the underlying laser's output with a much lower frequency—one derived from Lincoln Lab's radar work—that the system would actually use to image warheads. The specs were so extreme that outside contractors balked, so the team had to build the modulator themselves. This highly stable beam, now filigreed with a modulated signal, would then go to the amplifier Edwards's team had better make work.

The amplifier was to work much like an acoustic amp working on an old record player. Those of us old enough (or, with the resurgence of vinyl, young enough) to be familiar with old-school turntables know that you can turn off the volume and, if you put your ear close enough to the needle, hear the tinny high pitches of an LP's music as the needle scrapes along the grooves. It was much the same story with the laser signal coming into the amplifier. This ultra-stable whisper of light had to be amplified into a full-throated shout, but with none of the distortion you hear when you turn up a stereo too loud.

Edwards and colleagues' creation would be called the Coherent Optical Radar Amplifier, or CORA. "We had very little imagination when it came to naming things," Edwards says. "Why couldn't it be called Apache or something?"

To generate extreme power with extreme fidelity, they ended up with a system weighing seventy tons and filling a forty-by-seventy-foot room (the Firepond laser system spilled across two buildings). The heart of CORA was made up of two roughly twelve-foot-long, eight-foot-tall gain modules whose stainless steel vessels were built by a company in Oregon that normally made reactors for nuclear submarines, Edwards says. Inside each would go a ten-foot electron gun. Each electron gun pulsed up to ten times a second, coaxing the gas inside to conduct electricity long enough for a separate system to push through thousands of amps of current in a flash lasting sixty-millionths of a second.[23] The electron guns generated enough X-rays that someone standing close to CORA would be dead in about eight hours, Edwards calculated, so he had what he describes as "these beautiful stainless steel modules" coated with a three-eighths-inch layer of lead. The result resembled boilers in the basement of an aging middle school, ignoring the cryocooler and the dozen orange and black hoses swooping down into them. Painted blue like the Little Engine That Could, they sat on thick steel rails a freight train could ride in on.[24]

**The Firepond CORA amplifier. The lead-coated gain modules are toward the back of the room from this vantage point. (Photo courtesy of MIT Lincoln Laboratory.)**

Surrounding these massive blue gain modules were racks and racks of electronics full of gauges and switches, and with them, wires and pipes and tubes and big canisters of gas. This was a metal-dominant clutter that, upon closer inspection, was actually quite deliberate. Still, the overwhelming impression was of a physics lab gone feral. CORA alone cost about $24 million, all told—about double that in today's dollars.[25]

Once properly amplified, the laser light bounced off a mirror at the bottom of the tower and up into a telescope we've already met: the forty-eight-incher Fiocco and Smullin used for Luna See. In 1971, it had found its way back to Massachusetts from its missile-hunting work in Maryland. It hosted three lidar systems now: the amplified, modulated Firepond laser; one with blue light to illuminate satellites or rockets for closer inspection by the main infrared beam; and a third with green light to do laser ranging on satellites with corner cubes. A couple of smaller telescopes for target acquisition or other uses, depending on the experiment, were mounted to it, as was a phased array radar disc to alert folks in the control room to aircraft straying into Firepond's three-million-watt pulses. On the receiving end, the Firepond team had to invent a digital signal processor to turn the Doppler lidar data into video a human could recognize. That took a computer developed for acoustic submarine detection. Its components filled an entire wall of a room. Another wall was full of computers to keep a thirty-foot-wide laser beam pegged to a fast-moving object hundreds of miles away.[26]

These are just some examples of Firepond's immensity and complexity. As the five-year program wore on, the Firepond team faced increasing skepticism that they could get it all to work. Until they could turn it on, they could only argue that physics and design were sound. The entire Star Wars program increasingly faced doubts from scientists, Congress, and the press.[27] And so the pressure mounted to produce results.

Finally, in March 1990, Firepond system tracked its first satellite—Seasat, the radar altimeter that paved the way for ICESat. Twenty-five days later would come the first big test. A sounding rocket would launch from NASA Wallops and eject a canister that would inflate into a six-foot cone. From more than four hundred miles away, the laser was to watch it all unfold.

The amplifier, the weak link in the old Firepond system, was in the hot seat. Rather than watch from the control room, Edwards went to the "engine room" with the team working the amplifier. He donned a headset that blocked out the noise of the room and piped in the mission control audio. Without hearing protection, the bullwhip-cracking of the electron guns would make you dizzy, they had found out from experience. CORA seemed to be running fine. They had tracked and imaged plenty of satellites in the intervening three weeks. No need to worry, he told himself. Through the headset came "3-2-1 . . . Wallops has a launch."

This was the sounding rocket. They now had millions of dollars in the air. Edwards felt himself go cold. If the laser broke, it was on him.

It would be two minutes until the rocket climbed high enough to start lidar operations. Then another two until the canister emerged and the cone inflated. He closed his eyes and listened to the amplifier as it fired, willing it to keep going. He had a sense of what combination of guns and pumps and motors sounded normal and what didn't, even through the ear protection.

The voice over the intercom offered updates: "Beam blocker up" and "Target ejection on my mark. Mark!" and "We see separation."

There was quiet. Then Edwards heard oohing and aahing, and a "Look at that!" and a "The Force is with us" and an "It's beautiful." And then it was over, and the intercom erupted in cheers. Five years after he had started making the thing that had better work, he felt no desire to cheer—only relief.

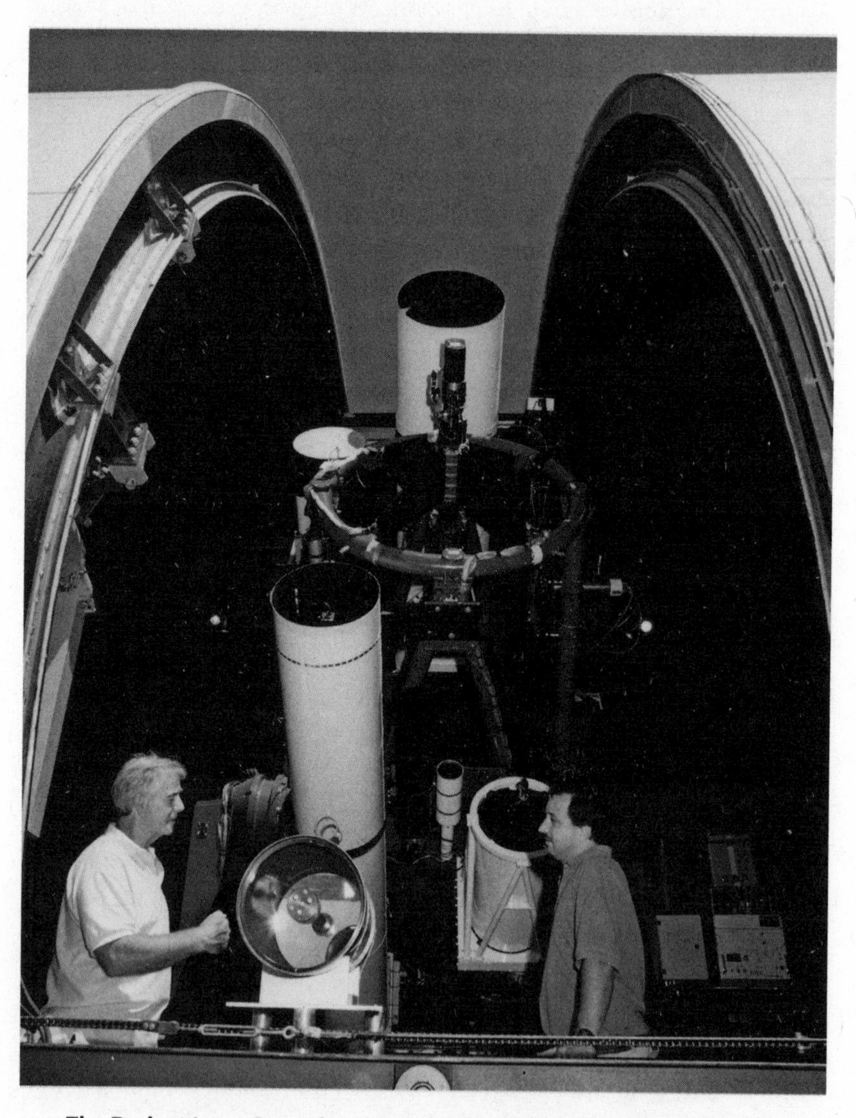

**The Project Luna See telescope, as adapted and augmented for use at Firepond. (Photo courtesy of MIT Lincoln Laboratory.)**

In the coming months and years, the Firepond team followed up three more big rocket experiments and tracked many satellites, the old disco ball in space LAGEOS among them. And then, in March 1993, Firepond was snuffed out.

A couple of things had changed. The SDI concept evolved from what people increasingly recognized as the truly impossible task of shooting thousands of missiles and warheads out of the sky to one more focused on shorter-range missiles in conflict zones. The Strategic Defense Initiative Organization became the Ballistic Missile Defense Organization that year (in 2002, it became the Missile Defense Agency).

Also obvious was that the gargantuan Firepond system could not be shrunk into anything close to an airborne, much less spaceborne, system. Firepond was a technological masterwork, but one that, for the foreseeable future, had no path forward.

Plus, in 1991, the Soviet Union had collapsed.

"I thought I was going to spend the rest of my career at Firepond, but sometimes world events have other ideas," Edwards says.

Firepond had cost perhaps $100 million. It sat unused for a few years until it was disassembled entirely in 2000. Only the old Luna See telescope remained. It's used for satellite tracking still.

But here's the irony. Firepond was among the few shining examples of the United States' immense investment in SDI bearing tangible, if experimental, fruit. And SDI scared the hell out of the Soviets. The big agenda item at the 1986 Reagan-Gorbachev summit in Reykjavik had been SDI, which could subvert the logic of mutually assured destruction. Critically, it would force the Soviets to build their own such system, which Gorbachev knew his ossifying republic could ill afford—they were spending 30–40 percent of GDP on defense already (the US figure was closer to 6 percent).[28]

Reagan refused to budge on SDI, something the *Washington Post* derided with the headline "Nonexistent Weapons Undid Summit." Gorbachev pushed ahead more quickly than he otherwise might have with a hodgepodge of political and economic reforms that collectively might generate the revenue required to pay for a Russian SDI equivalent. It was too much for the system to bear. And so, as one analyst put it, "SDI was the straw that broke the Communist camel's back."[29]

The Soviet Union was collapsing by the time that CNN story aired in

March 1990. Still, the Doppler-lidar-fed video of that expanding dummy warhead certainly reached Kremlin eyes, underscoring the need to somehow keep fueling their own missile defense research despite running on fumes. Had that been Firepond's sole contribution, it just might have been enough. But a tiny corner of the Firepond project lived on too—and would lead to an entirely new way of seeing the world.

# CHAPTER 12
# MAP QUEST

Richard Marino joined the Firepond project in 1985, fresh from his experimental particle physics PhD at Case Western Reserve, a few miles from where he had grown up in Maple Heights, Ohio. He abruptly went from working with a particle beam of antiprotons (to see how they reacted with protons, in search of what he describes as "an unusual state of matter") to helping develop the greatest lidar the world had ever seen.

He spent his time creating computer models of how well Firepond could be expected to see and what the pictures created from the massive system's data might look like. As Firepond took shape, he worked on tests for it. But he recognized, too, that any Doppler system remotely approaching Firepond's muscle would be too big to fly. To get a lidar close enough to an incoming nuclear weapon that it could help dispatch a killer beam or projectile in time, it had to fly. The leaders of the Strategic Defense Initiative were interested in trying other approaches too. Given his numerical modeling savvy, Marino was charged with simulating the performance of smaller lidars proposed by various contractors—in particular, lidars that might handle target discrimination from the nose of an antimissile weapon.

In the meantime, he was working on his own project. It had been sparked by what could be confused for a thought experiment: he wondered how little light energy you could get back from a lidar's target and still be able to tell one object from another. It was less esoteric than it seems.

If you can make do with dribbles of laser energy and still distinguish a warhead from a decoy, you can build a smaller, more efficient system. Or a lidar of the same size, weight, and power can see farther. That would give the interceptor carrying it more time to adjust its aim. The logical extreme would be a lidar that could do it by detecting individual photons, the smallest parcel of light the universe doles out. It happened that single-photon detectors existed: hair-trigger photomultiplier tubes of the sort John Degnan used for his satellite laser ranging. But they would be bulky and power-thirsty on a missile. Another possibility was something called a Geiger-mode avalanche photodiode. The thump of a single photon triggered an avalanche of electrons. They were originally used in gamma ray detectors in places like nuclear power plants.

Marino took interest in these detectors. Because they saturated quickly, they could relate that light was hitting them—just not how much. (This was different than, say, a tree-sampling full-waveform lidar, which could also capture the intensity of the return.) But there was a way around that, Marino saw. You'd be sending and receiving many, many pulses quickly with a Geiger-mode system. With each single photon received, you would know the time of flight, and thus how far the photon that triggered the cascade had traveled. With enough of those measurements, you could estimate intensity.

In addition, if each Geiger-mode pixel had its own timing circuit, you could build an array of them. You could split the outgoing laser into beamlets, one for each detector in your array. Each of those subdetectors would look only for returns from its designated beamlet. With the same thing happening across an array, each scattered pulse would yield a 3D picture of whatever it was you were trying to discriminate. So it would be really fast, and the laser wouldn't have to scan over the area of interest. Plus, if you only needed one photon back per Geiger subdetector, you could get by with a much, much smaller laser than Firepond's.

There was a catch. Stray light from about anywhere sent Geiger-mode detectors into a tizzy. They were really, really noisy. Too noisy for them to work with lidar. That was the conventional wisdom.

"The intuition is that if you're picking up hundreds of thousands of stray photons per second, you won't be able to find the ones you want. It would just be a mess to sort through," Marino says.[1]

But there were ways to cut through the mess, he knew. One was through an ability to measure time more precisely than any other quantity in the universe. As with the MOLA Mars lidar, you could tell the system to ignore returns outside of a brief time window in which the target should probably show up. A half-million stray photons per second becomes half a stray photon (statistically) per microsecond and less than one-thousandth of a stray photon per nanosecond. Then you can look for the particular color your laser fired and also focus in a very specific direction. The noise would become manageable, Marino calculated.

"Math prevailed when intuition failed," Marino says.

He looked for Geiger-mode detectors. He found a company in nearby Watertown, Massachusetts, called Radiation Monitoring Devices. It specialized in X-ray detectors, but one of its scientists, Stefan Vasile, had published a paper on a low-noise Geiger-mode detector he had come up with. Vasile was kind enough to let Marino borrow it, and Marino took images of a plastic model of an F-14 fighter plane and other simple objects in his lab while studying its potential. The noise was indeed manageable, and a system built around it would be much simpler than what was happening with the big Firepond laser.

Marino reached out to Paul McManamon of the Air Force Research Laboratory's Sensors Directorate in Dayton, Ohio. McManamon, in addition to his optical phased array work, was a driving force behind the military's interest in a lidar small enough to guide an antimissile interceptor. McManamon paid for another detector for Marino. It came from the same former RCA Canada lab that had made all of Goddard's space lidar detectors. Marino's earlier lab work showed Geiger-mode lidar to be promising enough that a photon-counting add-on was included in the last of Firepond's sounding rocket experiments, called Firebird 1B, in April 1992.[2]

With Firepond's cancellation, staffing in the group working on it dropped from seventy to thirteen. Marino went to a different Lincoln Lab division. He managed to keep his Geiger-mode work alive on internal funding, moving it forward slowly. Then he got married and, in 1995, headed to Lincoln Labs' facility on Kwajalein Island. The work would be interesting. Plus, "I thought, wow, we could go to the South Pacific on a two-year honeymoon," he says. Two years as a technical liaison in Washington, DC, followed the Kwajalein stint.

Before leaving for the tropics, Marino brought another group up to speed on his work and urged them to build on it. By the time Marino returned, Lincoln Lab was making its own thirty-two-pixel-square Geiger-mode detector arrays, complete with tiny timing circuits glommed onto the back of individual Geiger-mode pixels no thicker than a sheet of copier paper. The combination enabled sub-nanosecond timing for each pixel and range resolution of about three inches.[3] Geiger-mode imaging detectors were a reality.

Which leaves us with the laser. It would have to pulse quickly in addition to being small and light. There were options, but they tended to be bulky, complex, finicky, and expensive, which defeated the purpose of the entire Geiger-mode enterprise.[4] It happens that a Lincoln Lab scientist was working on one that fit the bill.

John Zayhowski had joined Lincoln Lab the year after Marino did, straight out of an MIT PhD program. He went to work on microchip lasers. These were blocks of lasing crystals with mirrored surfaces on opposite ends, much like Theodore Maiman's ruby laser. These were pumped not with a flashlamp but rather a diode laser. The early ones, made in 1987 and 1988, emitted a steady, continuous wave. They were tiny: about a cubic millimeter square, not including the diode laser pump. Over the next couple of years, Zayhowski got them to pulse with help from some extra electronics. The pulsed systems were more complex than he would have liked, though.

A Lincoln Lab project looking for better ways to clean up toxic Superfund sites came knocking in early 1993. They wanted a laser they could build into a cone penetrator to be pushed maybe a hundred feet into the ground. The laser would light up the toxics at different depths for spectroscopic analysis. It had to be small and tough, so the external pulsing electronics would have to go.

Zayhowski calculated that if he were to bond a second tiny crystal to the tiny crystal that produced a microchip laser's continuous beam, the resulting device could pulse in bursts shorter than a nanosecond, thousands of times per second, all on its own. The effect would be something like that of a huge bucket way up high in a water park, filling from a pipe until top-heavy enough to dump its contents on the squealing kids below, then repeating the cycle. Such a thing would be not only mechanically simple but powerful: the briefness of the pulses would make for flashes about ten thousand times more potent than the steady flow of energy the diode laser poured into it. For this environmental monitoring project, Zayhowski invented the passively Q-switched microchip laser.[5]

Zayhowski worked with MIT scientist Richard Heinrichs's group to use the new microchip laser as the source for the Geiger-mode detectors Marino had envisioned and Heinrichs's group had advanced.[6] The combination worked well, and by 1999, Marino was back at Lincoln Lab working on a program to move Geiger-mode lidar out of the lab and into the field, as well as help private industry get up to speed on producing it (DARPA, recognizing its potential, wanted to expand supply). Harris Corporation, a company known for its geospatial mapping prowess, and Princeton Lightwave and Boeing Spectrolab, optical detector specialists, became the chosen ones.

The first field tests, under a DARPA-funded, Marino-led program called Jigsaw in 2001, were designed to piece together partial 3D images of objects under forest canopies, if for rather different reasons than the ones motivating archaeologists in Central America. DARPA wanted a lidar small enough to fly on a drone. In parallel, the technology went into a higher-altitude 3D mapping system called ALIRT. That system used a different kind of Geiger-mode detector, one capable of sensing infrared laser light.[7]

With DARPA support, the Air Force Research Laboratory built HALOE, its own Geiger-mode lidar. By 2010, HALOE and ALIRT were mapping the entirety of Afghanistan, flying miles higher and gathering data ten times faster than traditional airborne lidar could, thereby enabling better urban warfare planning, line-of-sight analysis, and battlefield visualization, among other benefits.[8] That same year, ALIRT was dispatched to Haiti to map earthquake damage for disaster responders. ALIRT led to a string of descendants with increasing capabilities, and progress continues at Lincoln Lab, most of it in secret.[9]

Geiger-mode isn't the end of the single-photon story in mapping. John Degnan's single-photon lidar started with his quest for a better way to do satellite laser ranging. In 1998, he was looking at data from what would become his Next Generation Satellite Laser Ranging system. He noticed that if the satellite showed up a little bit earlier or later than expected, the system produced a slope in data. As Degnan studied this, the thought struck him: if I can see a slope from a satellite, why can't I see a real slope?[10]

Degnan pitched the idea to NASA Goddard's Instrument Incubator Program and by January 2001 was flying the NASA Microlaser Altimeter. The name fit: its microchip laser was just 2.3 millimeters long—less than half the length of that "2.3" you just read. Green like the SLR lasers and extraordinarily sensitive, the instrument scanned terrain and could see into the water for bathymetric profiles, in daylight, from as high as twenty-three thousand feet despite the tiny, low-power laser.[11]

In January 2003, Degnan retired from NASA after a thirty-eight-year career launched with Beacon Explorer B. He joined Sigma Space, where with an Air Force grant he designed a second-generation single-photon lidar, called Leafcutter, to fit in the nose of an unmanned aerial vehicle. This one included a GPS receiver and an inertial measurement unit and weighed just seventy-three pounds. It split the laser beam into a square pattern of one hundred beamlets matched to a ten-by-ten detector array, so a laser pulsing twenty-two thousand times a second yielded up to 2.2 million surface measurements in that short span. NASA tested it over land, forests, and ice from 2004 to 2007.[12] Degnan and Sigma Space followed up with other research systems—smaller for lower altitude, bigger ones to fly higher. By 2012, the Army Geospatial Center had added Sigma Space's High Altitude Lidar to its BuckEye program for providing soldiers a 3D ground-level perspective of urban areas and supply routes for mission planning and visualization, reconnaissance, training, and line-of-sight analysis.[13] Before long, Geiger-mode and single-photon lidar systems would shake up the commercial mapping business.

A year after my visit to the International Lidar Mapping Forum in Denver, I am back. I fancy myself a veteran this time. But I soon find myself gawking at various hardware in amazement. The $250,000 Leica Geosystems backpack isn't to be seen, but Austrian Microsoft subsidiary Vexcel Imaging has its UltraCam Panther on a mannequin's shoulders. It combines a compact, spinning lidar; a GPS unit; and, sticking up high over it all, a white ball with enough holes to accommodate twenty-six cameras. The cameras stitch together a 360-degree panorama.

Those interested in something less ambitious might go with a unit not much larger than a sportscaster's microphone, except the microphone is a Hokuyo lidar grabbing forty-three thousand distance measurements per second. The lidar, a squat cylinder a couple of inches across, has been mounted on a mechanism pinwheeling it once every two seconds. It connects to some electronics in a small shoulder bag. Del Stewart, who works for GeoSLAM, tells me this is his company's ZEB-REVO RT. Earlier, he was walking around with the thing, scanning the Hyatt Regency lobby. They've been selling the $45,000 devices since 2013, largely to customers in surveying and engineering for use in tunnels, mines, power stations, and buildings, Stewart tells me.

Teledyne Optech is here, having just announced the latest version of its ALTM airborne lidar, called the Galaxy Prime. It's a sixty-pound steel-gray box the size of something in which Amazon might send you a basketball. It includes far-evolved versions of about everything that once filled the NASA Wallops P-3 aircraft. It manages a million laser pulses per second, juggling several in the air at a time. It measures elevation to within eight inches from fifteen thousand feet, 30 percent higher than the ALTM's previous ceiling; from five hundred feet up, it will get you to within about an inch, vertically. There are other features with names like PulseTRAK and SwathTRAK aimed at making things easier on mappers. There's an optional digital camera so you can fuse the data into a three-dimensional, true-color, fly-through world, a common addition to high-end lidar. Optech has also brought along its Maverick mobile mapper, which mounts on a vehicle, and its tripod-based Polaris station. Missing is the Titan, whose three lasers at different wavelengths can see into the water and classify vegetation, among other feats. NCALM used that one to find the sixty thousand Maya structures in Guatemala.

Optech's competitors are right there with it. A couple of examples: Leica Geosystems, which made the aerial mapper that archaeologists flew over Cambodia, has just announced its TerrainMapper, which fires two million pulses per second. That's the same pulse frequency as the Riegl's VQ-1560i, the latest version of the scanner on NASA's Airborne Snow Observatory. And those are just the flagship, million-dollar machines.

Over in Riegl's display space, I note the new VMX-2HA, a mobile mapper with two lidars and seven cameras. The cameras peer out the ends of adjustable extensions that lend to a general impression of what might emerge if an outboard motor and a deep-sea crab got together. This particular creature can measure surroundings as far as a quarter mile away to an accuracy of two-tenths of an inch, from the roof of a moving car.

Riegl's backstory sort of rhymes with Optech's, but with Austrian physicist Johannes Riegl in place of Canadian physicist Allan Carswell. Riegl made his own radio receivers as a boy; as a PhD student at the Vienna University of Technology in 1968, he experimented with short-pulse semiconductor lasers. The semiconductor diode lasers that hit the market around 1970 seemed promising to him as a basis for compact lidars. By 1975, the Austrian company Eumig was selling a hydrographic surveying instrument based on his technology. The instrument didn't look into the water but rather back toward land, to establish the location of a sonar ship (remember, no GPS back then). Three years later, a university colleague suggested, "Johannes, your technology and your results are quite fine. But there is no scientific value and no future in it—better to leave the academic world and start your own business." Riegl launched his company—full name: Dr. Johannes Riegl, Radartechnik und Elektrooptik—in 1978.[14]

The name has gotten shorter as the product line has broadened. Riegl's early strength was in laser ranging. By 1980, Riegl sold a lidar for profiling tunnels; in 1982 came handheld "laser binoculars"; in 1985, a sniper rifle scope. With the early 1990s came pocket-size laser binoculars for hunters and a laser speed gun for traffic cops. Riegl continued on with industrial distance and speed sensors and, in 1996, took to the air with a scanning lidar for street and rail corridor mapping. The company got into stationary scanners on tripods in the late 1990s, mobile mapping in the 2000s, and, most recently, products specialized for drones.

Johannes Riegl Jr. is among the team on hand in Denver. Aside from his youth and having more in-control hair, the president of Riegl USA is a dead ringer for his dad. We have corresponded in recent months, during which he related a lot of the information in the preceding two paragraphs. I tell him I am particularly interested in one of the company's new airborne lidar units. It's smaller than a Kleenex box, weighs about five pounds, and shoots one hundred thousand laser pulses per second. It's made for a drone.

Drones are one of two recent seismic shifts in commercial airborne mapping. They enable, as Riegl tells me, a new way to do aerial scanning over agricultural fields, over forests, in complex corridors, along power lines and "in other areas that are hard or tricky to access with a typical surveying platform like mobile scanning or terrestrial scanning," he says. "So of course you can fly all these areas at high altitude and get a certain point density. But if you want certain smaller areas in an extremely high level of detail, then most likely UAVs will be able to fill this gap."

They're also a lot cheaper than big helicopter- and airplane-mounted systems, he adds. "It's become an entry-level driver, and they're building a totally new market." He walks me over to a much smaller booth in a crowded corner of the exhibit hall. It's what the International Lidar Mapping Forum organizers call the "UAV Pavilion." Riegl introduces me to Grayson Omans, CEO of Los Angeles–based Phoenix Lidar Systems. The company and others like it—YellowScan is directly across the way, and LiDAR USA has a booth around the corner—fills a market niche that didn't exist a few years ago. Phoenix Lidar takes lightweight hardware from the likes of Riegl and integrates it into drones or onto rack-mounted systems for a car or boat. It then sells the combination to companies selling 3D mapping services to companies that need 3D maps. He and colleagues brought along a big DJI six-rotor drone and an evil-looking Vapor 55 helicopter, both with little lidars mounted to their bellies.

Omans was a product manager for big solar companies when he recognized an opportunity in aerial lidar mapping in 2013. They focused on Australia and South America until the US Federal Aviation Administration released its Small UAS Rule in August 2016, which established guidelines for piloting commercial drones.[15] Now Phoenix Lidar is busy in this country, he says. As end users' needs get more and more specific, so do the platforms his company and others put together to solve them, he says.

"You're going to see a lot of companies cater towards specific applications—power line mapping to mining to you name it," Omans says. "If they want to map gas lines, now you have gas-integrated sensors."

Indeed, not a hundred feet away, a company called Microdrones is selling a quadcopter with a fully integrated methane inspection package. Chuck Anderson, chief strategy officer at Environmental Consultants and an expert in lidar for utilities, tells me later that there are still hurdles with respect to these unmanned aerial systems. The FAA limits them to fifty-five pounds, which affects range, and they can't as of yet be legally operated out of the operator's sight. They're also not cheap—$100,000 to $200,000 or more, depending on the instruments you fly. Anderson says his company is working with a utility to provide data to the FAA showing they can safely fly a single lidar-carrying drone down miles of power line, handing off control from one operator to another.[16]

The other seismic shift in commercial airborne lidar evolved from the work of MIT Lincoln Laboratory's Richard Marino and NASA/Sigma Space's John Degnan. At this conference in 2015, Harris Corporation introduced its Geiger-mode lidar; at the 2017 show, Leica Geosystems introduced a single-photon mapping lidar called the SPL 100. The pitch was similar for them both: you could scan at 330 miles per hour in a jet at 25,000 feet rather than at 110 miles per hour in a Cessna at 3,300 feet, collecting data ten to fifteen times faster with the same resolution or better than what suddenly became known as linear-mode lidar.[17]

As far as how these technologies moved from NASA and the Department of Defense: Hexagon already owned Leica Geosystems when it bought Sigma Space in February 2016, and as mentioned, Harris Corporation had worked on Geiger-mode lidar with MIT Lincoln Laboratory since the Jigsaw program in 2001.

Even as the aerial mapping industry wrestles with the implications of these powerful new entrants, the likes of Optech and Riegl have made a point of boosting the operating ceiling, speed, and point resolution of their linear-mode scanners. In trade journals and at conferences like this, titles like "Geiger-Mode/Single-Photon vs. Linear Scanning" have become mainstays.

**Downtown Seattle in November 2017, as captured by a Harris Geiger-mode lidar from an elevation of twelve thousand feet flying at 270 miles per hour. The color version shows things higher up, such as the Space Needle at right, in red hues, transitioning to greens and blues with decreasing elevation. (Photo courtesy of Harris Corporation.)**

I wander into a panel with this title. It features high-end people from Harris, Leica Geosystems, Riegl, and Teledyne Optech, among others. They seem to agree they all have a role to play. The single-photon and Geiger-mode systems covering massive areas will be great for big customers interested in big scans, such as the government of North Carolina or perhaps a utility looking for a purview of all its infrastructure. But linear-mode lidar has its advantages too, in terms of faster turnaround time and an ability to fly under low clouds and dispatch on short notice, to far-off places, and in tight weather windows because there are many more of these systems available to take the call. There seems to be some consensus that these new machines from Harris and Leica Geosystems are best suited to regional- or even continental-scale collections, whereas linear-mode excels in mapping the corridor and city scale. But nobody really knows where all this is going.

The end customers don't seem to care how their mapping data show up as long as the information is good and as inexpensive as possible. Alvan Karlin, a geographic information systems expert at the Southwest Florida Water Management District encompassing the Tampa-St. Petersburg area, tells the room, "Whatever the most cost-effective technology that gets us the product we need is the route we'll go."

A water manager in southwestern Florida uses lidar differently than one in the Rio Grande Basin. Karlin's sixteen-county, ten-thousand-square-mile district is concerned with water supply, flood protection, water quality, and the state of natural systems, he says. Lidar's been important in making sense of what Karlin describes as "Florida's deranged terrain," one bereft of deeply cut basins, with topography riddled with soluble limestone sinkholes between which lazy rivers form. Flooding is the overriding focus, it seems: Karlin shows aerial images of a subdivision overlaid with lidar-fed estimates of where ponds might form after a storm. It's used to make sure catchments do their job and don't send water over roads, and it's saved a lot of engineering time over the years, he says.

Hope Morgan, a manager with North Carolina Emergency Management specializing in geographic information systems, then talks about her state's work with lidar. Flood management is a big focus there too. When you think about flooding in the Carolinas, hurricane storm surges tend to be top-of-mind. But, Morgan says, "riverine flooding is actually the issue. We have a pretty good handle on what's happening with hurricanes." Her state ranks only behind Florida in terms of flood damage costs, she says.[18]

In 2000, North Carolina started mapping the entire state with lidar. It took until 2005, and it was, Morgan says, "one of the best ideas they've ever had." Thanks to more targeted field surveying, North Carolina saved $7 for every dollar spent on lidar, she says.

Subsequent lidar collections included 5.25 million structures, enabling more exact flood risk assessments for real estate across the state. Lidar data provided an elevation map upon which to base photogrammetry (which now also goes by the in-vogue term "structure from motion"). One-quarter of North Carolina—twenty-five counties—is flown with cameras every year, capturing land-use changes in particular (photogrammetry can also do digital elevation mapping where the ground isn't obscured).

Hurricane Sandy in 2013 brought in money for more flights, and in 2014, another forty North Carolina counties were flown with higher-resolution lidar, a project wrapping up, data-wise, in 2018. The new lidar data have already made a difference on the ground, she says. Combining the lidar data with river gauges helped engineers build what Morgan calls "inundation libraries," which showed emergency managers where the flooding was going to happen and in what order. During the torrential rains after Hurricane Matthew's landfall in 2016, those libraries led to the evacuation of a hospital and a jail, among other places, because the inundation models showed they would flood. It also predicted what dams would overflow and what roads would get cut off as a result. The lidar-based model, combined with ominous stream gauge readings, showed that the town of Princeville would seriously flood. Emergency managers evacuated it. It seriously flooded.

Next up, Morgan says, is to use the lidar data as the basis for a statewide inventory to see which bridges will emerge from floodwaters first. That will help prioritize and speed up the post-flood inspections needed before a bridge can reopen to traffic, she says.

Morgan and others sprinkle in the jargon "QL-1" and "QL-2" as they speak. These refer to quality levels for lidar data. Jason Stoker, the chief elevation scientist for the US Geological Survey's National Geospatial Program, was a driving force behind the terms becoming vernacular. He happens to be sitting right in front of me. A few months earlier, I sat across from him in his office in Fort Collins.

A schefflera plant takes over the corner of Stoker's USGS Fort Collins Science Center office, which, north-facing, is dusky despite the bright September day outside. The left armrest of his green office chair has gone missing, save for the metal support. He's in his early forties, longish hair parted in the middle, with a passing resemblance to the actor Kurt Russell.

Stoker's work is a modern twist on that of the US Geological Survey's most famous alum, John Wesley Powell. Despite having lost not an armrest but an actual arm in the Battle of Shiloh, Powell led a team down the Colorado River in 1869, mapping as he went. In 1881, he became the USGS's second chief. His goal from the start was to create a single national topographic map, one that would define the playing field for the American experiment. "A Government cannot do any scientific work of more value to the people at large than by causing the construction of proper topographic maps of the country," Powell told Congress in 1884.[19] It would be done one quadrangle at a time.

A quadrangle—strictly speaking, a rectangle or a square—was a topo map, a roughly eighteen-by-twenty-one-inch sheet hand-drawn back at the office based on a field team's calculation of angles and sketching of features based on such tools as alidades, plane tables, steel tape, and thirty-inch-long mercury barometers (these tended to break in the field, so teams brought extra glass tubes and a flask of mercury). Over the decades, the tools got better: aerial photography came in the 1930s, stereophotography in the 1940s, photogrammetry in the 1950s, electronic surveying instruments in the 1970s. But the basic approach to mapmaking stayed consistent—a human being interpreted inputs and produced a handcrafted map. One hundred and ten years after Powell kicked off the effort, the USGS finished up a thirty-three-million-person-hour effort (more than fourteen thousand work years) to create fifty-seven thousand topographic quadrangles covering the continental United States.[20]

The USGS later came up with software to translate all those hand-drawn topo contours into digital elevation data, creating a foundation for what they called the National Map. Stoker recognized that the National Map was only as good as the photogrammetry of the quadrangles it was largely based on. These were accurate to maybe four and a half feet vertically; a good lidar scan should get to within four inches vertically. By the early 2000s, lidar scans were happening in North Carolina, the Pacific Northwest, and many other places. Why not upgrade the basic elevation data in the National Map with much more accurate lidar data?

Such a program could pay for itself, he and colleagues calculated. They wrote up a report advocating for a national 3D Elevation Program, or 3DEP. It listed twenty-seven realms in which better elevation data could be bankable. Topping the list: flood risk management, in which a few inches of elevation can be the difference between basements staying dry and a deluge of insurance claims. Other areas included infrastructure and construction management, natural resource conservation, water supply and water quality, and agriculture and precision farming (slope dictates runoff, and therefore water, fertilizer, pesticide, and other needs). The list also encompassed areas of management including wildfires; volcanoes, earthquakes, landslides, and other geological hazards; and forests, rivers and streams, coastal zones, energy resources, and more. With $150 million in annual funding, most of which would go toward matching programs for state and local governments contracting out for lidar flights, Stoker's team estimated they could cover the three million square miles of continental United States in eight years. They calculated annual benefits of at least $1.1 billion a year by doing so.[21]

Better flood risk management would more than cover the cost of the lidar program, with an estimated $300 million in annual benefits. The Federal Emergency Management Agency is partnering with the 3DEP program as it updates the maps upon which the National Flood Insurance Program is based. (That program, which covers five million households, is, as the *New York Times* put it, "broke" from the combined burden of various hurricanes and Superstorm Sandy.[22])

"A lot of their maps were originally done, like ours were, in the seventies," Stoker says.

But it's clear in talking to Stoker that the value of such maps would also come from unexpected corners. Construction companies could use the lay of the land to better estimate how many truckloads of dirt a cut-and-fill operation to put in a road might take, saving trips. Data on water depth of lakes and rivers can help with everything from navigation to drought management to keeping fishermen happy. Data on road slopes could help smart cars accelerate slightly at the end of a downhill to save gas on the subsequent uphill. Data on forests would help climate scientists estimate carbon balance. Lidar's ability to see through the canopy could help save lives and property from landslides of the sort that killed forty-three and wiped out forty-nine homes in Oso, Washington, in 2014. 3DEP is, in fact, a key part of a federal bill that would create a National Landslide Hazards Reduction Program.[23]

3DEP is where the terms QL-1 and QL-2 (for "quality level") came from. QL-1 calls for four-inch vertical accuracy and eight lidar pulses per square meter. QL-2 requires the same accuracy but with two pulses per square meter. 3DEP requires QL-2 data. The QLs descend to QL-5, which involves high-altitude radar that the program proposed for Alaska, which is too huge, too empty, and too cloudy to cost-effectively cover with lidar.[24]

By late 2017, 3DEP had covered about 40 percent of the country, Stoker says. With the current $80 million annual budget, it'll take more like fifteen years to finish than the initially proposed eight. Then, ideally, they would rescan to capture forest growth and destruction, urban development, and other changes, much like Landsat satellites have done using imagery.

"So it's exciting times for a government guy, you know?" Stoker says. "There's so much potential out there that it'll be cool to see how much of it actually gets realized by either us or the private sector."

John Wesley Powell would surely agree.

# CHAPTER 13
# ATOMS TO BYTES

The Riegl and Teledyne Optech mobile mappers on display at the Denver conference are far from alone: there's also Vexcel Imaging's UltraCam Mustang, Leica Geosystems' Pegasus:Two, and Trimble's MX9, which looked like something off the top of the *Back to the Future* DeLorean. Mitsubishi Electric, Isan Technology, Topcon, 3D Laser Mapping and Dynascan, among others, also sell lidar-plus-camera systems. Their abundance is a hint of their usefulness in building true-color 3D renderings of streetscapes, rail corridors, construction sites, mines, and more. The winding path that led to the growing business of mobile mapping is a story in itself, one that started somewhere it's not advisable to drive: in the bowels of nuclear power plants. Or, more accurately, in the mind of Ben Kacyra, who had made a good part of his living from that inhospitable setting.

Kacyra, with a head of dark curly hair, a hard-to-place accent, and soft-spoken charm, was a civil engineer. He arrived in the United States from his native Iraq in 1965, earning a master's degree in civil engineering at the University of Illinois at Urbana-Champaign in a year in which he subsisted disproportionately on McDonald's burgers. He got a job offer in Cincinnati, Ohio. He had never heard of the place, but he bought a map and drove down to work mostly on designing highway bridges. He met a woman who worked in the accounting department, Barbara. In December 1968, Kacyra, twenty-seven, hitched a U-Haul trailer to his '65 Ford Mustang and drove west to the San Francisco Bay Area, where he intended to try his luck. In the passenger seat was Barbara.[1]

Kacyra got a job with an engineering firm, where he worked to design office buildings and hospitals using longhand calculations and a slide rule. He urged his boss to buy time on the University of California, Berkeley's mainframe. When the boss declined, he opened an account for himself. Rather than parking in front of the television after dinner, the Kacyras drove over to Berkeley. He coded, Barbara worked the punch cards, and their baby daughter slept in a wicker basket. He saw opportunity in applying computational power to the problem of understanding how a building might respond to an earthquake—no airy hypothetical in a seismic zone like San Francisco. In 1973, he cofounded Cygna Engineering Services, which became a pioneer in 3D structural analysis and dynamic earthquake analysis.[2]

Cygna grew. Nuclear power plants became a specialty, more so after the 1979 Three Mile Island accident. At these plants and elsewhere, Kacyra ran into the same problem. Before they could do an accurate computer simulation, they had to understand exactly how the structure had been built. You might think blueprints would take care of that. But changes happen during the construction process: the owner, last minute, wants two rooms where there was only one, or the blueprints have a pipe running through a beam. This happened a lot—enough so that construction companies were generally expected to provide "as-built" drawings incorporating the changes. But they were an afterthought, done poorly if at all, Kacyra says.

So Kacyra sent Cygna teams of as many as forty people in with tape measures, ladders, clipboards, and, often, radiation suits. It was dangerous, imprecise work with the door wide open for all sorts of human error. There had to be a better way, Kacyra thought. But this way worked well enough to grow Cygna to five hundred employees and, in 1989, attract an offer from Kaiser Engineers. Kacyra, forty-nine, sold the company. His immigrant-making-his-fortune-in-America dream fulfilled, he might have just coasted into early retirement. But this as-built problem nagged at him. He wished there were a way, as he puts it, "to go into this nuclear power plant and have a box with you and you just set it there and start the camera and—zip—you have a CAD drawing. That was the dream.

"I wanted this," Kacyra continues, tapping a pillar next to the table at which he sits, "inside of my computer. Atoms to bytes."[3]

He parsed the dream into ambitious specifications that, fulfilled, would bring the sorts of improvements in accuracy and efficiency the market couldn't ignore. He settled upon being able to measure a space to an accuracy of a millimeter from a distance of one hundred meters (the thickness of a fingernail from a football field away, roughly) with something battery powered, affordable, and rugged enough to handle the construction business. He had no idea what technologies might be able to make it all happen. He called Jerry Dimsdale, whom he thought might.

Dimsdale was a PhD engineer from UC Berkeley who had stuck around to redesign and then run Berkeley's twenty-by-twenty-foot earthquake simulator, a task that required high-end skills in areas ranging from differential equations to computer programming to hydraulics. Kacyra, who had worked with Dimsdale and been impressed, thought of him as a "gizmologist" who could "build a gizmo to do it one way or another." Kacyra explained what he was after and wondered if cameras or radar might work; Dimsdale felt that lasers were probably the only way to get that sort of accuracy. Neither of them knew much about lasers. So Kacyra cold called Robert Byer of Stanford University, one of the world's top laser physicists. Dimsdale and Kacyra drove over to Palo Alto and explained their idea. Byer told them they wouldn't be breaking the laws of physics, but they would be pushing the envelope.[4]

"And I thought, 'Super! We'll go do it!'" Kacyra says. "Little did I know what he really meant by 'pushing the envelope.'"

Byer offered another suggestion. It would take an extremely fast-pulsing, simple laser to do this thing. They might consider reaching out to John Zayhowski at MIT Lincoln Laboratory.

In early 1993, Dimsdale and Kacyra visited Zayhowski in Massachusetts. He had, at this point, just started working on the passively Q-switched microchip laser for the Superfund cleanup project. Zayhowski made clear that one-millimeter accuracy at one hundred meters would be, in this sort of package, beyond the laws of physics. They settled on six millimeters (about a quarter inch) at fifty meters, which Kacyra believed he could sell—among other reasons, because he would have bought one for his own engineering firm had it been available. He worked out an agreement through which Zayhowski would help figure out what sorts of lasers and receivers might work for the system Kacyra envisioned and, if he found something, design and construct a prototype. There was no guarantee he would find anything, Zayhowski says. But, he says, "we were familiar with the state of the art and did have confidence in our ability to push it." He soon realized the new microchip laser he was developing might lend itself to the very different job of 3D scanning. Dimsdale spent five months at Lincoln Lab working with Zayhowski and developing the software for the system.[5]

That left the receiver—in particular, its timing circuit. It had to be extraordinarily fast—able to slice time into intervals shorter than seven picoseconds—trillionths of a second—to achieve millimeter accuracy. Dimsdale looked around and came upon work done at Los Alamos National Laboratories. The team there had developed something capable of the picosecond timing needed to watch an underground nuclear test unfold. The signal coming from the detonation was to arrive just ahead of the wire melting, one of the Los Alamos guys told Kacyra. The timing circuit sat on a shelf, unused. Kacyra signed another cooperative research agreement and, with a good deal of work again involving Dimsdale, went through three versions of the circuit before it was fast enough.[6]

In the meantime, Barbara had named her husband's new venture Cyra Technologies and its proposed product Cyrax, a privilege she would pay for, over the next several years, by running day-to-day operations and writing checks from her and Ben's account totaling several million dollars. The prototype arrived from MIT. It was on a one-by-three-foot stainless steel breadboard. They mounted it to another steel breadboard about twice that size and added hardware including the orange head of a surveyor's total station, which took individual laser shots to confirm distance measurements.[7] It was 1994, and it looked like something stolen from a physics lab. Two people could lift it, with effort.

They tested it on Barbara's marble statue of a horse, a section of I beam, four-inch PVC pipes, and, out in the parking lot of the small office in Orinda, Kacyra's Mercedes 230 SL coupe. For the car's scan, the prototype was hoisted atop a rack the size of a small refrigerator holding its supporting hardware. The system converted atoms into bytes in the form of what Kacrya and the half dozen engineers now working with him took to calling a "cloud of points." This was perhaps the first use of what would become the term of art "point cloud."

There remained a chasm between the working prototype and something Kacyra could sell. The team realized, for example, that the scan mirrors expanded and contracted with temperature changes. They would have to calibrate across a range of possible temperatures and tweak the firmware, which grew to hundreds of thousands of lines of code. Just as pressing was turning the flood of data coming off the prototype into something an engineer could use—that is, a CAD drawing.

Kacyra bought a high-end Silicon Graphics machine and assumed that the mainstay AutoCAD would just work. It went into the computational equivalent of cardiac arrest at ten thousand or so points (the prototype's scan of a sheet of copier paper half a football field away might have a million points). He reached out to UC Berkeley computer science professor Carlo Sequin. Sequin suggested that Kacyra didn't have a CAD problem but a visualization problem. Kacyra drew up yet another agreement. Sequin's grad students worked on two things. One was a way to see what the scanner was producing in real time; otherwise, users might not realize they missed something until they were back at the office. The Berkeley students also developed software to parse a point cloud and calculate, based on a pixel's relationship in space to the ones near it, whether it was part of a plane or a cylinder or a box—the sorts of shapes a CAD program (and an engineer) can work with. The combination led to what would become Cyclone, the world's first integrated point cloud software. As with so much else with this enterprise, Kacyra had had no idea his fledgling company would have to invent it.

By 1997, Kacyra was looking for someone with deeper pockets to help turn their Alpha, as they called the prototype, into something they could take to market. It had been five years since he had first suggested the idea to Dimsdale. "I mean, we were dumping so much money into it, I had to get investors," he says. Chevron had been on his short list of possible strategic partners—oil and gas fields and refineries are constantly changing.

To clear his head and work his body, Kacyra rode his bike. He was taking a break during a ride up Mount Diablo east of Orinda when another cyclist stopped to do the same. They struck up a conversation. Steve Brown turned out to work for a Chevron group looking for ways to boost productivity. One meeting led to the next, and soon came the idea of a test at Chevron's Richmond refinery on San Francisco Bay.

Dimsdale was opposed. This was still an experimental system bolted to a metal board the size of a kitchen table. It wasn't ready. The company's future depended on it, Kacyra insisted. That and they would need to use Dimsdale's VW Vanagon, which had more room for what Dimsdale later described as "our ponderous contraption" than Kacyra's Chevy Suburban. Dimsdale relented.[8]

For the test, Chevron chose a roughly twenty-foot-tall cooling tower on a fifty-square-foot pad. This was a huge radiator on thick I-beam stilts, augmented with compressors and snaking pipes and handrails and supports, the whole thing painted black—the least reflective color in the universe. If their ponderous contraption worked here, it would work about anywhere. But would it work?

They set up the prototype in the Vanagon so it could shoot the laser out the sliding side door. The computer hardware was in the back so they could observe the monitors through the open rear hatch. Dimsdale had to jack up the right side of the van to get the aim right. As they scanned, a black limousine arrived. Out stepped Chevron vice chairman Jim Sullivan, unannounced—the sort of guy who could move millions of dollars with a nod of his head.

The Cyra team moved the van around the cooling tower to capture it from different angles. The raw feed on the screen sketched the tower in an eerie green, as if glowing algae had self-assembled the image. Some minutes later, Cyra's custom software had converted the whole thing into a CAD-digestible file, which a CAD program running on a laptop computer was happy to display. Sullivan checked out the monitors and laptop screen and was, Kacyra recalls, "taken aback. You could see it."

Chevron became a strategic partner, as did the engineering firm Fluor Daniel and the US Navy. By 1998, the Cyra team had packaged their science experiment into a formidable black box. It was big, but it fit on a thick tripod Kacyra poached from his ten-inch Meade telescope. The supporting electronics went from the size of a small fridge to that of a dorm fridge. The Beta, they called it, until they called it the Cyrax 2400. It was the world's first tripod-mounted, commercial 3D scanner.[9] Over the next couple of years, Cyrax machines scanned Chevron facilities, the USS Tarawa, a Detroit Edison power plant, a water utility concrete pour, Highway 880 before a widening project, a bridge in Pennsylvania, caves in Oregon Caves National Monument and Preserve, the movie sets for *Starship Troopers* and *Moulin Rouge* (for more realistic special effects), and much more. About the only thing they didn't scan was a nuclear power plant.[10]

Among the early adopters was Marcello Balzani of the University of Ferrara in Italy. He applied a late-breaking Cyrax feature—mapping not only the location of a returning point but the intensity of the signal coming back from it—to monitor the degradation of Italian ruins. Kacyra's team had discovered the capability when they realized their lidar could read the license plate of a software engineer's Pontiac Firebird. Historical preservationists scanned the Roman Colosseum; the temple at Chavín de Huántar, Peru; and the ruins of Tikal, Guatemala.

**Ben Kacyra (left) with Cyra Technologies electronics engineer Kevin Heppell at the site of a new ramp off Interstate 680 in Walnut Creek, California. They were testing the Cyrax 2400's ability to digitally capture rebar placement in freeway columns. (Photo courtesy of Ben Kacyra.)**

Kacyra had set out to solve the narrow problem of improving as-built design engineering. He ended up inventing the terrestrial 3D scanning industry. Leica Geosystems CEO Hans Hess heard about the system and ultimately flew in from Zurich to see it for himself. Leica bought the company in 2001; its HDS tripod-mounted scanner line shares the Cyrax as a common ancestor. Cyclone software is still part of the package. Riegl, Optech, Trimble, FARO, and many others sell competitive systems far and wide.

Kacyra was now sixty-one and might have been reasonably expected to kick back and enjoy the fruits of serial entrepreneurship. But, as he puts it, "I think the way I grew up, I felt that people are here on this Earth to do something good, and not just do one thing and walk away. You're here to continue to contribute. I mean, I truly believe that."

He and Barbara established the Kacyra Family Foundation. It would invest in improving medical care and helping schools in disadvantaged areas. Also, it would invest in historic preservation, an interest of both Kacyras, in Ben's case since he was a young boy in Mosul. His father, a diplomat turned entrepreneur, was an archaeology buff, a habit sustained though a library of books young Ben paged through with interest, as well as through trips to nearby sites such as the ancient Assyrian city of Nineveh. It was a couple of miles beyond the Tigris River, on whose banks the family lived. They had picnics under the watch of the winged bulls carved into the gates of the city.

Kacyra imagined historical preservation as a sidelight, perhaps sustained with an occasional lecture to spread the word about the Cyrax scanner's ability to capture accurate 3D renderings of historically significant places. But a month after Leica's acquisition of Cyra Technologies was announced, Afghanistan's Taliban destroyed the Buddhas of Bamiyan: hundred-plus-foot-tall, 1,500-year-old statues carved into a sandstone cliff 140 miles northwest of Kabul. His technology could have at least preserved the site digitally and provided a detailed model for the Buddhas' rebuilding. Now they were just gone.

He explained to the new foundation's board that he had talked to people in the cultural and heritage preservation world. There would be no mad rush to buy Cyrax machines—they were expensive, they were complex, and preservationists didn't think they needed them. "We're going to have to go out ourselves and do it and show them," he told the board.[11]

The foundation helped fund projects using Cyrax scanners, including Balzani's work in Pompeii and UC Berkeley archaeologist Alonzo Addison's work in Tambo Colorado, Peru. By 2003, it was time to spin off the historic preservation work into a nonprofit of its own. The team would do their own projects to show the world how powerful the technology could be. Then others would do their own projects. The new nonprofit would then serve as a clearinghouse for 3D data of historic sites all over the world.

John Ristevski, a Berkeley grad student who had spent time with Addison in Peru, was involved with Kacyra's architectural-preservation venture early. As an undergraduate at the University of Melbourne, Ristevski had used surveying equipment and Hasselblad cameras to capture the ancient Thai capital of Ayutthaya and ancient Olympia in Greece. He had met Addison at a conference in Japan, where the Melbourne team was presenting their Greek work. Addison suggested Ristevski come to Berkeley for graduate school, which he did, under Inca-archaeology expert Jean-Pierre Protzen. One of Ristevski's first classes in 2001 involved a Cyrax scanner. By 2003, Ristevski had been on multiple expeditions to Peru, and also to Angkor in Cambodia with the Cyrax 2500, the second and last commercial model produced under the Cyrax name. Ristevski's PhD research focused on 3D documentation of buildings for heritage preservation. An advisor told him about Kacyra's new nonprofit idea, which had the same goal.

"I went over to Orinda and had a chat with him," Ristevski says. "And this was right up my alley—just what I wanted to do."[12]

Ristevski and Kacyra worked through the initial concepts for the organization that summer. Ristevski came up with a name: CyArk, for "Cyber Archive." From 2004 to 2006, he worked with CyArk as he worked on his PhD. CyArk did its own scanning and also collected scans from others in the field, bringing in digital versions of dozens of sites around the world. In 2006, Ristevski left both CyArk and UC Berkeley. He and Anthony Fassero, a CyArk project manager, cofounded a company they called earthmine (all lowercase). CyArk was the inspiration, Ristevski says. CyArk was doing what Kacyra nicknamed "scan and can." CyArk posted projects to its website, but the datasets were so huge—two thousand to four thousand gigabytes—that the bulk of it was simply stored for posterity. That information, Ristevski says, "was amazing—super-dense and superaccurate. But could you actually use it?"

The idea behind earthmine was to invent a way to do large-scale 3D data capture and then make it usable in solving real-world problems. The focus would be streetscapes. Given that lidar scanners like the Cyrax were both pricey and not suited to trundling along on a roof rack, earthmine looked to other technologies for the data capture. Ristevski knew photogrammetry well from his undergrad days in Melbourne. He looked around for something that might work on a vehicle. He found it at the NASA JPL.

The Mars Spirit and Opportunity rovers flew with a couple of camera systems. The Pancam, which sent home the pretty pictures, got all the attention. Ristevski was interested in the Hazcam. It used stereo-vision-derived photogrammetry to calculate distances to hazards as the rover navigated.

Without the Hazcam, there would have been far fewer pretty pictures. With the rovers now roving the red planet, the technology was, as Ristevski puts it, "on a shelf at Caltech," the California Institute of Technology, which manages JPL.

They struck a deal with JPL to boost the Hazcam's resolution as earthmine developed software to automate the data's path from the car's roof to the cloud to the end user. Such a system could inventory streetscapes—utility poles, signs, garbage cans, parking meters, benches, building facades, the pavement itself. It could help track billions of dollars in assets for real estate companies, property assessors, transportation and public works departments, and others, Ristevski and Fassero believed. It might even provide a basis for augmented reality.

The earthmine system captured hundreds of thousands of miles of streetscapes over the next few years. Google was doing the same thing. Working with Stanford professor Sebastian Thrun and Anthony Levandowski, the company sharply reduced the cost of its Street View mobile mapping systems. Topcon built the hardware based on designs by a startup called 510 Systems, which Levandowski had founded shortly after joining Thrun at Google.[13] We'll see those names again.

These threads converged into what became the mobile mapping industry, spawning the hardware I mentioned at the top of this chapter as well as proprietary systems that Google (which bought 510 Systems in 2011) and others are using to turn atoms to bytes for the consumption of self-driving cars. The effort also became the basis for the street-level views in Google Maps and Microsoft's Bing Maps.[14]

Ristevski and Fassero sold earthmine to Nokia in 2012; Nokia had already bought Navteq. With Ristevski leading the effort, Nokia added a lidar from Velodyne, a company we'll be talking more about, and became Here, which Nokia sold to a group of German automakers in 2015. By then, Here mobile mapping systems had rolled in thirty-five countries, capturing millions of miles of roadway each year.

Ristevski stayed on until May 2016. It had been a decade since he worked with CyArk. Soon, he would be back.

High on a wall against which two bicycles lean in CyArk's Oakland, California, offices is a thick strip of corkboard. Tacked to it are all sorts of pictures, many of them extracted from magazines featuring CyArk and its work.

Words, numbers, and scribbles fill the walls below, except for the wall with the backpacks, climbing harnesses, cables, and a big tripod hanging from it. A world map on another wall hosts pins marking the locations of the two hundred sites in forty countries on seven continents CyArk and its partners have scanned as of late 2017 (or as of whenever someone in a busy ten-person operation last had time to further puncture it). Elizabeth Lee, CyArk's vice president of programs and development, greets me and guides us to a conference room. On the walls, there are poster-size CyArk clips from the *New York Times*, *Archaeology*, *National Geographic*, the *San Francisco Chronicle*, and *American Surveyor*. That last one's a cover with Ben Kacyra in a leather jacket, arms crossed, the words "Monumental Challenge" encroaching on his chest. Lee has started into a presentation when Ristevski stops in to say hello. Together, they explain CyArk's new direction.

Kacyra had approached Ristevski about joining the CyArk board right around the time he left Here. Kacyra hinted that he was thinking about stepping back from the day-to-day work to make way for someone who could lead CyArk for years to come. He was, after all, seventy-six now. Ristevski saw an organization that had come a long way and was still doing the work that had attracted him all those years ago. In November 2016, Kacyra stepped down (he and Barbara remained on the board) and passed the chairman/CEO torch to Ristevski. He would deliver infinitely more value than his annual salary, which would be, at his request, zero. (Fassero came back to CyArk too: he's now vice president of capture and production.)

They are expanding on the original vision of scanning, canning, and sharing lidar data. Preservation for potential restoration was still a focus, and one that had already come in handy. CyArk scanned the Kasubi Tombs UNESCO World Heritage Site in Uganda in 2009. The next year, a fire destroyed it. They sent over the data for its reconstruction. In Myanmar, a scan of a temple in Bagan helped managers understand how the site had changed after a 2016 earthquake. Maps generated from CyArk scans have provided a basis for site managers' day-to-day operations. In partnership with the International Council on Monuments and Sites, CyArk trained Syrians how to use a tripod lidar scanner and a drone with a camera. Despite the ongoing civil war, they have scanned six sites, mostly in the historical core of Damascus, and they spirit hard drives with terabytes of data out of the country via hand courier, Lee says.

That's all still going to happen, Ristevski says. But in addition, they want to broaden access to these sites in new ways. "One thing we found is that trying to just save heritage digitally, it's hard to get people excited about that mission. If you pull the right heartstrings, maybe they'll care about it a little bit. But a lot of people just don't know enough about these places to care about them," Ristevski says. CyArk intends to take them there virtually, at which point it's easier to say, "Hey, you've gone here virtually, you've seen these places, they're amazing, they're also at risk. Help support our mission to collect them, preserve them, archive them, and share them," he says. Among the ways they're doing it is through a partnership with Google Arts and Culture called Open Heritage, which launched in April 2018. It shares 3D renderings and also the raw lidar and photogrammetry data from more than two dozen sites, with more to come.[15]

Kacyra arrives wearing the same leather jacket he has on in the *American Surveyor* cover on the wall. His hair is just a bit grayer. As Lee continues her presentation, he asks questions and adds perspective. She explains the combination of tripod-based lidar scans and photography, drone-based aerial photography, and audio recording that's being combined into building "hyperrealistic environments." We walk to a nook outside the server room back near the entrance. There's a small desk with a computer. On the wall is a big photo of two guys in yellow hard hats dangling with their laser scanner next to Abe Lincoln's nose on Mount Rushmore, and a second one of Kacyra doing his 2011 TED talk (at the end of the talk, he laser scanned the audience and displayed the results on the big screen a few seconds later).

Kacyra dons an Oculus Rift VR headset connected to the computer. Lee explains how to navigate using a controller he holds in his hand. And moments later, he is among the temples of Ayutthaya, the ancient Thai capital. The rendering he sees is from a small subset of the data from 170 lidar scans and ninety thousand photos, 2.7 terabytes in all, captured by two CyArk people in a week. Lee guides him as he teleports among the strange towers. He turns as he looks around, wrapping himself temporarily in the long connecting cable, interjecting his virtual tour with the occasional "Oh, this is great" and "Whoa."

"You know who would really like this? Barbara. I gotta show this to her," he says.

He extricates himself from the headset, and I spend a couple of minutes in Thailand, and then closer to home, in a kiva cliff dwelling on a sunny day in Mesa Verde National Park, Colorado. This is still "a very early prototype," Lee says. It adds audio and an augmented reality layer over the virtual reality—things to click on and have a park ranger explain. I turn toward the ledge. Kacyra, looking at some of what I see on a computer monitor, says, "Step toward the ledge," which I do. The hollow sound of the wind adds to the sense of space. If not quite vertigo-inducing, it's *real*. Which I can say with at least some authority, my atoms having been there a couple of years ago.

Kacyra is enthusiastic about the new direction. "These monuments here—the temples—why were they built? What went on there? Taking off my hat as an engineer, an entrepreneur, an inventor, I'm saying, really the most important thing is for our children to learn the story. That's really what we can leave them for the future, is the story. And it's hard to tell the story without this."

Kacyra and I leave for a long lunch across the street, during which he tells me his story. As we part ways, I can't imagine him slowing down much in this second retirement.

CyArk pressed ahead with its virtual reality work in the months after my visit. In February 2018, the nonprofit launched its MasterWorks VR program, offering up Mesa Verde, Ayutthaya, Mount Rushmore, and Chavín de Huántar for virtual reality visitation via free Oculus Rift and Samsung Gear VR downloads. I'm guessing Ben has shown Barbara, and I'm quite sure she liked it.

Despite having been founded with Leica Geosystems money, CyArk uses laser scanners from FARO Technologies, another mainstay of commercial lidar.[16] Optech and Riegl started in laser physics; FARO started in surgical glue.

Simon Raab was a PhD student working in a basement lab at McGill University in Montreal. It was 1978, and he was at McGill because he couldn't afford the tuition to continue at Cornell, where he had earned a master's degree in engineering physics. At Cornell, he had studied the properties of aluminum crystals, building his own electron microscope in the process. At McGill, for his mechanical engineering doctorate, he focused on biomechanics and biomaterials. The problem of fixing joint implants attracted him. This was a surface physics problem of getting metal implants to stick to a plastic cement. Often, it went wrong, inviting infection and complications. Raab came up with a Plexiglas-like epoxy and patented it. "That money, whatever tiny royalty, allowed me to found FARO," Raab says.[17]

He founded FARO with Gregory Fraser, another PhD in the basement lab. Fraser was working on a tool to precisely measure how the ankle moves—its kinematics. He built a small device for this purpose. Raab programmed a calculator to deal with the outputs and also helped secure test articles availed through motorcycle accidents and diabetic amputations, transporting, as he puts it, "some poor bugger's leg in a taxi."

They called the company Res-Tec when they launched in 1981. While they focused on developing software-aided devices to help with diagnostics and surgery, they also did R&D on things as disparate as nonflammable fire hoses and packaging problems. By 1983, they had gotten tired of working on other people's problems, changed the company's name to FARO, and focused on medical hardware and software. Their first product, in 1983, was the Genucom, *genu-* being the Latin root for knee. It was designed to quantify the laxity of an injured knee to millimeter accuracy. Raab built the Genucom's computer around an early Intel 8086 processor and added a green-screen front end. The Metrecom, for analyzing spine problems, followed in 1987. The next year, FARO released the Surgicom, designed to work in concert with medical scanners to help brain and spine surgeons locate difficult-to-find tumors and remove them with minimal collateral damage.[18] General Electric, which made medical scanners, was a customer, until it suddenly wasn't. The orders stopped when GE decided to go with infrared cameras.

"I said, holy shit, we could be obsoleted by this optomechanical technology," Raab says. "That's the moment when FARO decided we'd better get into optics."

In 1990, Raab and Fraser moved the company to Lake Mary, Florida, to be closer to customers and Raab's family. Shortly after that came a strange call. It was from a quality manager at Martin Marietta. Martin (now Lockheed Martin) was a big defense contractor. The guy on the line, Daniel Buckles, was involved in aircraft and naval systems outside of Baltimore. This was a long way from the orthopedics docs, surgeons, chiropractors, and physical therapists comprising FARO's target market.

A Martin executive had seen FARO's 3D measurement arm in a mall, where a chiropractor was using it to recruit patients by measuring their back curvature, Buckles told Raab.

"How accurate is it?" Buckles asked.

"To a millimeter," Raab said.

"Any chance you can make it maybe twice as big and maybe ten times more accurate?"

"Sure," Raab said, though he wasn't.[19]

Buckles assigned a young engineer, Michael Raphael, to work with Raab on turning the medical device into a manufacturing tool. Raphael had been at Martin since graduating from Virginia Tech with an engineering science degree in 1985. His role was to solve manufacturing problems across programs, including ones making Patriot missiles, Titan rockets, and thrust reversers for General Electric and Pratt & Whitney jet engines (for slowing an airplane down after landing). This was in the days of mezzanines full of guys working on big drafting tables, when an ashtray was an office supply. With no internet, young engineers hopping between dispersed design and manufacturing centers acted as a project's connective tissue and immune system, holding things together and spotting problems before the rot spread too far.[20]

Raphael worked with Raab as FARO adapted the medical device to manufacturing and looked for ways to put it to use around Martin's factory near Baltimore. The technical name for it, an electrogoniometer, was not a help.

"I got jeered," Raphael says. "'Get that gonad-measurer away from me,' sort of thing. There were cartoons posted around the factory."

He found use for it with a big thrust reverser. These were, Raphael recalls in a conference room twenty-seven years later, "the size of that wall." They were cowls built around the engine, and they had to be shaped just right. Out of the mold, they were placed in a huge steel-framed jig, which played the role of a retainer in the mouth of a kid with her braces just off. Despite the strength of the jig, the thrust reversers were coming out very slightly skewed. Raphael said the problem was with the big jig shifting under the force. No way, the guys told him, the jig was fine. So he measured it with the electrogoniometer. Sure enough, the tension imparted by the thrust reverser's graphite skins pushed the jig about a millimeter out of place. A millimeter is a big deal in the aerospace industry, and in manufacturing in general, where, as Raab puts it, "a centimeter may as well be a mile."

FARO named the finished product the FaroArm. There was still no laser involved—you touched a little ball at the end of the articulated arm to whatever it was you were measuring. Raab and his team wrote an interface to AutoCAD, the engineering staple. He also reconsidered his market strategy. There were perhaps fifty thousand orthopedic surgeons, physical therapists, and chiropractors combined, plus insurance and regulatory hassles in the medical space. There were a million manufacturers. FARO licensed the Surgicom to Medtronic and Brainlab and never looked back.[21]

Later, they would add a laser scanner to the end of the FaroArm and grow their product line through internal development and acquisition. The FARO Focus scanner CyArk uses, one of several examples, evolved from technology acquired in the 2005 purchase of the German company iQvolution.[22]

Meanwhile, the post–Cold War "peace dividend" was squeezing Martin Marietta. Raphael took a voluntary layoff in 1995. He had twin three-year-old girls and told his wife he could get another job in a year if entrepreneurship didn't work out. He and his boss's boss, Richard Lee, bought a FaroArm and moved into a cramped office in Baltimore. Their business, Direct Dimensions, would be in converting atoms to bytes. If the item was small enough, they scanned it in the office; otherwise, they loaded the FaroArm into the back of a minivan. They did lots of quality control and reverse engineering projects, much of it for military hardware that had been designed before the computer age. "Black Hawk helicopters weren't in a computer. Most legacy aircraft weren't, either," Raphael says.

His company happened into an interesting new market in the late 1990s. He was scanning eighteen-foot-scale model ships used in flow tanks at the Naval Surface Warfare Center when one of the guys there mentioned that a cleaning person almost knocked a bronze bust of Admiral George W. Melville off its pedestal the night before. "They were thinking, 'What if something happened to Admiral Melville? That would really suck,'" Raphael says. The fine detail would make it tough to scan through contact with the FaroArm's little white ball. Raphael wondered if they couldn't put a laser scanner on the end of it. Conversations with FARO and the French company Kreon Technologies led to hardware; the Quebec City company InnovMetric Software was already working on a package called PolyWorks to turn a point cloud into a triangle mesh for manipulation and 3D printing. Several months later, the Navy guys had their virtual Melville backup. They 3D printed an actual mini-Melville in celebration.

"If we had Melville brought in today, it'd be done before breakfast, and way better than what we did then," Raphael tells me in the conference room. It's in Direct Dimensions' Owings Mills, Maryland, facility. Its shelves are busy with 3D prints of a tiny fraction of all the other things Direct Dimensions has scanned over the years: various versions of former Baltimore Ravens linebacker Ray Lewis's skull-capped head, models of Sierra Nevada's Dream Chaser spacecraft, miniature buildings, oddly shaped and supercomplex metal components, a life-size bust of George Washington in a colorful checkerboard pattern.

Raphael, fifty-four, never did go back to a traditional engineering job. Like Kacyra, he possesses an air of enthusiasm and avuncularity not always associated with the field in which he trained. He has grown Direct Dimensions from the two-man operation to a thirty-person company with clients in manufacturing, construction, engineering, the arts,[23] and medical areas. Among thousands of other things, his company has scanned the Lincoln Memorial (with a Cyrax machine, back in 2001), the Liberty Bell, the Iwo Jima Memorial, Tomb of the Unknown Soldier, the propellers of the Wright Flyer and an Israeli navy boat, the Washington National Cathedral, Times Square (for the film *The Amazing Spider-Man 2*), and the original Nike Air Jordan basketball shoe.

"The guy brought it in a briefcase handcuffed to his wrist," Raphael says.

He walks me to a work area where Boma Jack, a Direct Dimensions 3D artist, guides a FaroArm with two hands to track a laser line across the target's ornate metalwork. "It's some sort of historic light fixture, and the company makes reproductions," Raphael explains.

Lidar is only part of the story now, though it remains important in industrial applications demanding high precision and with big stuff—cars, trucks, boats, aircraft, and buildings. As we leave Boma Jack to her laser scanning, Raphael points out the old FARO Metrecom that started it all, folded into its carrier's waffle foam. He shows me a photogrammetry booth where a hemisphere of twenty-four Canon digital SLRs interspersed with lights combine to digitize a person's face into something a special effects artist can keep out of the uncanny valley. There's an even bigger one down in Georgia, he says, where two Direct Dimensions employees have spent the better part of a year doing lidar and photogrammetry scans for the movie *Avengers: Infinity War.* They've scanned about everybody and everything in the Marvel movie, Raphael says, to no small degree because "it's a whole lot cheaper to blow things up in the computer than it is to do it in real life."

**Direct Dimensions 3D artist Boma Jack scans a light fixture with a FaroArm.**

Another big customer is artist Jeff Koons, who relies on exquisitely detailed laser and photogrammetric renderings—involving more than ten thousand photographs of fifty megapixels each—to capture items (inflatable bunnies, say) in such detail that he can turn them into big, flawless sculptures. Raphael walks past a section of propeller shaft for a navy ship toward the back door. Out back, trucks are parked so the cabins can be laser scanned by some competitor of the truck maker for ergonomic evaluation. Back inside, he shows me the prosthetics room. It seems straight from a dental office except for the paintbrushes; squirt bottles of pigment; and collection of synthetic ears, eyelids, and noses. A Johns Hopkins Hospital anaplastologist sees patients here to replace body parts destroyed by illness or injury. Direct Dimensions scans the subjects to create silicon facsimiles. Ears are the easiest, Raphael says: it used to be they'd take a cast of your good ear and manually sculpt the opposite one for hours. "Now we bring a scanner over and scan a person's ear in minutes. Then, in the computer, we hit the mirror button, and it's now an opposite."

I'm still trying to process it all as we wrap up, but I manage to ask

Raphael where he sees all this headed. He doesn't hesitate: virtual and augmented reality. "We had a HoloLens"—Microsoft's entry in the VR hardware space—"in here yesterday. If you had a 3D CAD model of renovations, you can see the changes—you're looking at walls that aren't actually there. Seeing pipes in those walls," Raphael says.

Not so different, really, than bringing some Kiva in Mesa Verde to the VR-geared-up masses.

I walk back out to my rental car. I'll be driving this one, but sometime soon, the technologies we've talked about will be a big reason why, sooner than later, its descendants will drive themselves.

# CHAPTER 14

# HIT THE ROAD (SLOWLY)

All the lidar we've talked about—to sniff the air, to probe the depths, to discern the shape of the Earth and the orbits of satellites, to see the forest from the trees, to map streets and snow and ice and planets, to make the physical digital at power plants and world heritage sites and on Hollywood sets—it's all behind the veil as far as our everyday lives. Lidar has been a tool of science and government and commerce. It has helped us understand our world, build on it, and make it safer. But besides golfers with their laser range finders and those of us who have gotten speeding tickets recently (radar guns are mostly lidar guns these days), lidar isn't an everyday thing. That's soon going to change.

Work on autonomous vehicles is revving up, and the vast majority of self-driving cars and trucks will include lidar as a pillar of the sensor suites that make them possible. Lidar has been pivotal to the emergence of vehicle autonomy. To understand the difference it has made, consider where self-driving cars came from.

The Stanford Cart got its bicycle wheels in 1960–1961, when mechanical engineering graduate student James Adams put it together to test whether NASA could remotely operate a moon rover with a television camera from Earth (the 2.5-second round-trip light time was a problem, he found). But the cart itself, with electric motors powered by a car battery, became a test bed for efforts of grad students in Stanford's Artificial Intelligence Laboratory, or SAIL, who adapted and advanced the Cart. By 1966, it could reasonably follow a bright white line painted on a road or athletic field.[1] By then, SRI, which did a lot more than atmospheric lidar research, had built a robot called Shakey (named after the teetering nature of its motion). This was an indoor machine tethered to a mainframe computer. Given an hour or so to contemplate its next move, it could plan, find routes, and rearrange simple objects.[2] This, like the Stanford Cart, was really an artificial intelligence effort that happened to be on wheels—something that hasn't changed since.

SAIL PhD student Hans Moravec took his turn with the Stanford Cart starting in 1971, adding a mechanism that captured a succession of images as it slid a television camera laterally. He did this after observing small lizards he had caught and kept in a terrarium. "The lizards caught flies by pouncing on them. Since flies are fast, this requires speed and 3D precision," Moravec wrote. "Each lizard had eyes on opposite sides of its head; the visual fields could not overlap significantly, ruling out stereo vision. But before a pounce, a lizard would fix an eye on its victim, and sway its head slowly from side to side. This seemed a sensible way to range." Moravec thereby introduced photogrammetry to the Cart, and with it, depth perception.[3]

The system worked the same way today's autonomous vehicles do. It asked itself two questions: 1) Where am I? and 2) What's nearby? Then, based on sensor input, it calculated its next move, moved, and repeated the process. Moravec's cart relied on a mainframe capable of 2.5 million instructions per second—a modern iPhone processor is 240,000 times faster, the difference between a mile away and the distance to the moon. It took fifteen minutes for the Cart to answer those questions and plot the next meter of its journey. It made it across an intentionally cluttered room in about five hours.[4]

In 1980, Moravec joined the Carnegie Mellon University faculty in Pittsburgh. Directly and indirectly, the nexus of autonomous-robot research moved east with him. Moravec and his grad students built new machines to expand on the work he started on the Stanford Cart. Pluto, Neptune, and Uranus, the first three were called. They made real progress: Neptune took just one-fifth the time to cross a room scattered with obstacles than the Cart had. But that was still an hour.[5]

Four decades removed, these robotic ventures may seem quaint. A Roomba vacuum cleaner can cover the better part of a house in less time, taking care to avoid furniture and the dog. Consider, though, that William L. "Red" Whittaker, a colleague of Moravec's at Carnegie Mellon, recognizes such work as having been "beyond the state-of-the-art."

"This is fundamental creation," he says.[6]

Whittaker would do his share of it. His approach would differ from Moravec's and that of SRI. He wanted to solve problems to improve the science of autonomy. But he really wanted to solve actual problems using autonomy. His vision was to build "robots that would develop, secure, feed the world, and explore worlds beyond." Long before the sensor and computing technologies needed to enable competent autonomous robots existed, he was thinking about who might use them: farmers, miners, logistics people, explorers, and, one day, carmakers.[7]

Many a dreamer had mused about such things. Whittaker was not a dreamer but a worker. His first real job growing up in Hollidaysburg, Pennsylvania, was hammering railroad spikes. His dad sold explosives to mining and highway construction firms, so he traveled a lot. His mom, who had become a pilot at Piper's plant in nearby Lock Haven during World War II, stayed at home with the kids but taught herself chemistry and later worked in the profession in steel and mining. Her eldest son was sharp enough to become one of the youngest registered surveyors in the state and eventually get into Princeton. Whittaker started there in 1966. When money got tight, he joined the Marines during the Vietnam War. They sent him not far from where his mom had flown Pipers, and he learned how to program a computer at Lock Haven University.[8]

Whittaker finished up his civil engineering degree at Princeton in 1973. He thought about going into computing but felt the field would explode with or without him. "And I wanted something where it would be by my hand, and I would be cause in the matter, and where it would fulfill in my time and would really change the world."[9]

His Carnegie Mellon civil engineering PhD focused on robotics. When he finished it in 1979, he stuck around at the Carnegie Mellon Robotics Institute. Raj Reddy, who had arrived from Stanford's SAIL lab a decade before, had founded the institute that year. Whittaker was already thinking bigger than the repetitive motion machines starting to show up on factory floors. The 1979 Three Mile Island nuclear power plant accident two hundred miles east of Pittsburgh, which led to a partial meltdown of one of its reactors, gave Whittaker an opportunity to build tethered remote-control machines equipped with video cameras; bright lights; and various tools to explore, sample, and ultimately help clean up the badly contaminated nuclear power plant's basement.[10]

Real-world autonomy, capable of doing real work outside (or in contaminated basements), not in labs, was his aim. But Whittaker faced a chicken-oregg problem. As of the early 1980s, the idea of autonomous robots actually doing something productive was so far-fetched you couldn't attract corporate or government research money. But to prove it was possible, you needed research money. He sat down with Reddy, his boss.

"Look, the real opportunity is outside," Whittaker told him. "And by the way, it'll work, and it'll mean a hundred times whatever this is going on in here"—he motioned toward the labs with robots named after outer planets—"and I want it, I'm up for it, I stand for it. I'll deliver it for a quarter million, and I'll get it inside a year, and it's going to do something."

Reddy had been thinking about a broader vision including autonomous land, sea, and air vehicles. He considered his intense young colleague: close-cropped red hair thinning up top, the light beard to match. Tall, muscular, ready for a fight. "Well," Reddy said, "How about six months and $50,000?"[11]

Whittaker convinced some students to work with him starting in May 1984. They had a rudimentary version of their Terrestrial Navigator, or Terregator, rolling by August. It wasn't much to look at: a thigh-high metal box, six feet by two and a half feet, riding on six small wheels. A gasoline-fed electric generator of the sort you'd find behind a carnival booth powered it. But it wasn't supposed to be pretty; it was supposed to be able to go over curbs, stairs, or small logs; deal with steep terrain; and skid-steer like a tank (though without the treads). Just as importantly, Terregator was a test bed. The gyroscope and inclinometers buried inside next to the radio transceiver would combine with two types of "specialized devices to enrich a vehicle with more capable control." Specifically: "visual sensing and acoustic-based navigation systems."[12]

These would provide the "sense" in the "sense-plan-move" behind all intelligent autonomous motion, including your own. The "visual sensing," a camera, would, obviously, see. The acoustic-based navigation system would detect distances without the overhead of a vision-recognition system like the one Moravec's Stanford Cart used. In fact, Moravec's Carnegie Mellon group had already combined a camera and an acoustic-based system on their Neptune robot. Using both, they believed, could help build "a more complex and rich description of the robot's environment."[13] Today, it's called sensor fusion.

The idea behind the acoustic-based system was the same as with shipboard sonar. It was a time-of-flight system using sound waves rather than lidar's light waves. These early robotic sensors could thank the autofocus sonar on high-end Polaroid instant cameras for their existence. The cameras used a poker-chip-size diaphragm that emitted an ultrasonic click, listened for it, and adjusted the focus based on distance calculated by time of flight. Polaroid also made industrial versions of these, called Polaroid ultrasonic transducers. Roboticists used them in a couple of ways.

One way was to arrange two dozen or so of them in an outward-facing ring. With it, you could render a 360-degree "view" of surroundings up to about thirty feet away perhaps once every ten seconds. These became known as Denning rings, after the Woburn, Massachusetts, company that made them, Denning Mobile Robotics. James Crowley, a Carnegie Mellon PhD student turned Robotics Institute professor, came up with a second way. It pointed a single Polaroid sensor up to a reflector that narrowed the sensor's fat (thirty-degree cone angle) beam and spun on an axis like a searchlight, pitching the sound in different directions, catching it again, and calculating distances.[14]

Terregator was, early on, outfitted with a video camera and a Denning ring. But sonar had its weaknesses. It was very short-range. Its fat beam angle made for big spots. That was fine if you happened to be perpendicular to a nearby wall. But what if you weren't? Think of the "spot" of sound as the focus of a flashlight at a dark campground. Point it straight down and you have a circle. Aim out at the ground in front of you as you walk and it's an oval, and the distance to the beginning of it and the end of it can be extreme. That's not ideal for a distance-measuring device. Whittaker looked for something better. He found it at the Environmental Research Institute of Michigan, or ERIM. It was a classic case of making your own luck.

As Whittaker was advocating for what would become Terregator, DARPA was ramping up its billion-dollar Strategic Computing Initiative. The focus was on artificial intelligence for defense uses ranging from autonomous vehicles to "personalized associates" and battle management systems.[15] Between Moravec's research, that of Carnegie Mellon computer scientist Takeo Kanade, and others', CMU was already in a strong position to vie for some of that money. Whittaker's Three Mile Island work strengthened the case; Terregator even more so. DARPA money flowed starting in January 1985. Some of it bought what became known as "the ERIM."[16]

This was an interesting device from an interesting company. ERIM the company had been a University of Michigan–operated defense research nonprofit called Willow Run Laboratories. The university spun it out in 1972 in the wake of Vietnam War protests. William "Bill" Brown, who had been

Willow Run Labs' director and became ERIM's founder, chose the name "basically to get students off their back."[17]

This was "a real spook shop," says Dwight Carlson, who cofounded Plymouth, Michigan–based Perceptron and was a longtime friend of Brown's.

"I mean, there were guys working at ERIM that Bill Brown, the president, didn't know what they were working on. It was need-to-know, and the government decided he didn't need to know. And he was running the place," Carlson says. "They were very, very innovative, very, very advanced, and had a lot of PhDs from around the world. It was a very creative place, and all military."[18]

ERIM was a top radar house and holography pioneer. The ERIM lidar emerged from military contracts starting in the mid-1970s. By 1977, they had a version they could fly and, from an aircraft, tell the difference between a tank and a picture of a tank. By the next year, they had used it on the ground, again to identify tanks. This was the first lidar to scan and identify objects (as opposed to identifying how far away an object was).[19] ERIM compressed its truck-bed- or airplane-filling lidar into an eighty-five-pound baby-blue box about the size of a microwave oven. DARPA bought three of them. One went to "a government agency," another to Martin Marietta, and the third to Carnegie Mellon.[20]

The ERIM was a "push broom" lidar, with a spinning mirror sweeping laser light horizontally and a second mirror nodding up and down. The combination could render things to an accuracy of about six inches as far away as sixty-four feet, and it filled a 256-by-64 pixel screen twice a second with what it scanned (a modern smartphone screen has that many pixels in a space smaller than your pinkie fingernail).[21]

From the user-friendliness perspective, the ERIM was similarly adrift from a modern smartphone. It slipped out of calibration, it overheated, it was hard to troubleshoot. But it doubled the range of the Polaroid sensor and shrank the cone angle to a half degree.

"By comparison to sonar, you can imagine what a revolution it was in this game to have something with the precision, the spot resolution, the refresh rate, and the data rate," Whittaker says.[22]

Terregator was crawling through an underground coal mine by 1985, but it was clear that so small a vehicle could only do so much. The computers of the day were overmatched no matter what. But having to radio transmit sensor and other data to a computer room for off-board processing, then transmit it back before each move, was a recipe for long stops between steps. A fair chunk of Carnegie Mellon's DARPA money went into an autonomous vehicle big enough to carry the computers. It was called Navlab.

**Carnegie Mellon University's Navlab 1, the granddaddy of today's self-driving vehicles. The ERIM lidar is on top of the cab. (Photo courtesy of Carnegie Mellon University.)**

This was a Chevy panel van that Whittaker's team of students and technologists stripped down to the undercarriage. They built it back up, replacing the gearbox with a two-speed hydraulic transmission (easier to control with a computer and able to move very slowly, very precisely); beefing up the suspension; and constructing an external shell for racks of computers, controllers, telemetry, and internal sensors (Carnegie Mellon computer science, electrical engineering, and other collaborators were involved too). They added a twenty-kilowatt generator—it had five times the output of Terregator's—to power all the add-ons. They painted the whole thing as sky blue as a UPS truck is brown. They added an RV air conditioner to the roof to keep all the hardware from cooking itself. The ERIM lidar rode on top of the Navlab's cab; next to it perched a video camera. Much of the work happened outside on a loading dock during the Pittsburgh winter.[23]

"I can reflect on wasting a good amount of my life modifying, altering, and scratch-building these kinds of devices to be amenable to computer control," Whittaker says. "It didn't matter if it was a car or a tractor, drive-by-wire, brake-by-wire, steer-by-wire—back in the day, you had to make it up."[24]

Navlab was street legal in manual mode, capable of a very noisy thirty miles per hour; autonomously, with the combination of the ERIM, the camera, a lot of fancy software, and an inertial navigation system like you'd find in an aircraft, it moved along at strolling speed. But the step-stop machine had become a continual-motion machine.

Sanjiv Singh got into robotics because as a University of Denver junior in 1982, he became mildly ill after eating a sandwich. He had accidentally gotten gasoline on his hands, and it got on the sandwich, which he ate anyway. This had happened because with his classwork and his work in a DU computer lab, he was busy to the point that he had resorted to eating a sandwich while pumping gas.

"So I thought, you know, at that moment—this is like, this engineer's brain, OK?—Why do I have to stand in the cold and pump the gas? Couldn't we do this with a robot or something mechanical?"[25]

He designed such a robot for one of his classes, becoming enamored of the geometries and the physics involved. That led to his discovery that robotics was emerging as an actual field. He applied to Carnegie Mellon for graduate school and was rebuffed; ended up at Lehigh University in Bethlehem, Pennsylvania; and earned a master's degree in mobile robotics. As he was finishing up in the spring of 1985, he tried Carnegie Mellon again, this time seeking a job as a technologist. He made a couple of phone calls and landed on Whittaker's line. With Navlab nigh, Whittaker was hiring, but he wanted someone right away; Singh had commitments at Lehigh until August. Whittaker said he'd have to think about it.

A few days later, Singh called back. Whittaker's secretary answered. Whittaker was on the phone with a journalist and was about to go to Russia that evening, she told him.

"I'll tell you what: I'm gonna stay on the phone for as long as it takes," Singh told her.

"Well, it might be an hour," she said.

"That's OK," Singh said. "I'll just stay on the phone. Just tell him I will be on the phone."

Forty-five minutes later, Whittaker picked up. Singh told him he still couldn't be there until August, but he'd still really like the job. Whittaker said, "Fine, just come in."

Singh first worked on Navlab. Between that and a project to develop autonomous haul trucks for strip mines, he got to know the ERIM and its follow-on Perceptron lidars (the Perceptron had better range and higher resolution and weighed about half as much as the ERIM).[26] That was right around the time Whittaker's group, which had been part of Carnegie Mellon's civil engineering department, changed its name to the Field Robotics Center and moved into an abandoned power generation plant on campus. "Imagine walking into an old steel mill and there's abandoned stuff, random office equipment and papers and junk all over the place," Singh says.

Singh soon had a new directive, one that emerged after Whittaker visited the lidar pioneer Johannes Riegl in the town of Horn, Austria. The setting was as low-tech as Whittaker's abandoned power plant. "There was a barn," Whittaker recalls. Still, Riegl the company was by then highly regarded in the civil engineering world for its laser ranging systems. Whittaker was looking for a spin on that.

He wanted a laser version of James Crowley's lighthouse-style sonar sensor. The laser light would shoot up out of a laser range finder, bounce off a mirror and out to a target, and capture reflected photons along the same path. Spin the mirror and you'd get a rotary line scan, one yielding a continuum of distance measurements through a slice of everything around you at whatever height the mirror was mounted.[27]

Johannes Riegl thought it was a perfectly good idea and came up with a price of about $30,000 for first-of-its-kind engineering capable of surviving on the nose of Navlab or similar outdoor robots. Whittaker figured his team could do it for less, so he only bought a laser range finder from Riegl.

"How hard can it be to spin a mirror?" Whittaker asked Singh, whose job it would be to build the mechanism to do so.

It can be really hard to spin a mirror, Singh would find out. Spinning things per se is not hard for a roboticist. But add the mirror and the laser, and now you're dealing with optics, which is to robotics what French is to English. There's some overlap, but you're not going to be reading Proust out of the gate.

In mid-1987, Singh started the project to create what would be called Cyclone. The Riegl RF-90 range finder was boxy, with two coffee-mug-size lenses and a handle on the bottom. It was accurate to less than four inches at 160 feet, so Singh's product promised to be an improvement over the ERIM. Singh removed the handle. He put the range finder in a bigger box and added a motor and drive belt to spin a small tower containing the mirror. He added separate boxes for power and communications with Navlab's computers. He built custom circuitry for all the data moving back and forth.[28]

With the laser pulsing 7,200 times a second against a mirror spinning up to 15 times a second, a big challenge was knowing exactly where the mirror was pointing with each pulse of the laser (unless you did, you wouldn't know what you were measuring). There were also surprises that added time and expense. The motor generated lots of electrical noise, interfering with the signals to and from Navlab. Regardless of what coatings Singh tried, the glass window through which the spinning mirror saw the world blocked the laser's light. He ended up just leaving it open when it ran.

It took eighteen months and cost about three times what they would have paid Riegl, and it was loud like a Cuisinart someone forgot to turn off. But it did ultimately work. They mounted it on Navlab. By then, with GPS becoming an option for navigation, the lidar's big contribution was obstacle avoidance. They also put it on another Field Robotics Center creation called the Locomotion Emulator. In an underground mine, Singh and colleagues' software matched Cyclone's lidar output with a preloaded map of the tunnel system on the fly, enabling the robot to navigate to an accuracy of four inches.[29] This sort of navigation foreshadowed Singh's very different lidar work years later.

Navlab became Navlab 1 when its first successor, a modified military Humvee, took to the road (and off-road). That one also used the ERIM and

Perceptron lidars. By 1995, a couple of Carnegie Mellon PhD students took Navlab 5, an old Pontiac Transport, across the country on Interstate 70. It drove itself 98 percent of the time without lidar: it had only a camera and an inertial measurement for guidance.[30]

But Whittaker and others on the CMU team continued to view lidar as an indispensable sensor for avoiding obstacles. GPS was great for staying on the right track in general, and cameras could do a lot too in terms of recognizing what was road and what was roadside. But lidar provided an instant grasp of how far away something was, and with minimal computing overhead. That was important because the Field Robotics Center's ambitions far outstripped what computers could support, even without a lot of vision processing overhead. Those ambitions manifested in all sorts of ways during the 1990s and early 2000s.

Ambler, designed to navigate steep, boulder-riddled terrain as a prototype for future planetary robots, was among the wildest of them. This NASA-funded behemoth was fifteen feet tall with three pile-driver-like legs on either flank. It moved forward by lifting a rear leg, rotating it to the front of the line, and dropping it on whatever happened to be there. Testing involved bringing several boulders into the former power plant's high bay. Whittaker's team used both the ERIM and Perceptron lidars with it.[31]

They also built a robot called Dante to rappel into Antarctica's Mount Erebus and sample its volcanic gases (that one was foiled by a snapped fiber-optic cable); a robot to autonomously mow hay and alfalfa; and robots to roam the interiors of fuel storage tanks, hunt for meteorites, roam desert environments, and do subterranean mapping and exploration.[32]

Whittaker's motivation for the last of these was the 2002 Quecreek Mine disaster in Somerset County, Pennsylvania. Nine miners were trapped for three days after a breach into the flooded tunnels of an abandoned mine—an all too common occurrence. Inaccurate maps were to blame. Whittaker saw an opportunity for a robot mapper. In sixty days, he had a graduate robotics class build a 1,600-pound mobile robot called Groundhog out of the front halves of two ATVs. They added a video camera and line-scanning lidars on either end, these made by the German company SICK, a new option at the time. These did much the same thing as Singh's Cyclone lidar but were rather quieter. Groundhog slogged through ankle-deep yellow muck in a stretch of abandoned mine. A Carnegie Mellon computer science professor, Sebastian

Thrun, developed something called Simultaneous Localization and Mapping (SLAM) software to help Groundhog map and explore at the same time.[33]

Two years later, Whittaker's group followed up with a more sophisticated version of Groundhog called Cave Crawler. That one, led by PhD student Aaron Morris, used SICK lidars too but rotated them like propellers, which turned the line scanners into 3D imagers.

As Cave Crawler came together for mapping the subterranean voids, Whittaker started building a much more ambitious autonomous vehicle, one to roam the desert, with eyes on a $1 million prize for doing it the fastest.

# CHAPTER 15
# DEBACLE IN THE DESERT

By the time Red Whittaker and his Carnegie Mellon team got their autonomous Humvee to the Mojave Desert for the 2004 DARPA Grand Challenge, they had shortened the sense-plan-move cycle of autonomous motion from the fifteen minutes or so of Moravec's Stanford Cart to about a quarter of a second. They would need every last bit of it.

The race would start the morning of March 13, 2004. Red Team, as the group from Pittsburgh called itself, was among fifteen finalists that had qualified at the California Speedway in Fontana the week before. The race was 142 miles of winding dirt road starting in Barstow, California, and finishing in Primm, Nevada, just across the state line. To finish before the ten-hour cutoff, the vehicles would have to average fifteen miles per hour and hit fifty on flat stretches. This was a vast leap from the five miles per hour of Whittaker's first autonomous Humvee, Navlab 2. But then, the intervening decade and a half of technological advancement, not to mention robot-building experience, promised to change the game.

DARPA had announced the prize a year before, the motivation being a sentence in the 2001 National Defense Authorization Act. It said a third of US military ground combat vehicles should be unmanned by 2015.[1] Military contractors were moving too slowly. DARPA officials hadn't known quite what to expect when DARPA Director Tony Tether dangled a $1 million cash prize to the winner of a driverless marathon across the desert, complete with boulders, drop-offs, twists, turns, and cacti. The demands were a huge leap beyond anything even Whittaker had attempted. Still, the DARPA people had been pleasantly surprised with the 106 responses they got, of which 25 had been invited to qualify.[2]

This was a Frankencar freak show, with driverless dune buggies; ATVs;

jeeps and SUVs; crawlers of the sort Whittaker would send into a mine; six-wheeled rovers; ruggedized utility carts; a motorcycle; and the truly formidable TerraMax, built on an Oshkosh military monster truck. Most of them used one or more of the same SICK lidars Whittaker had put on Cave Crawler. Sandstorm, Red Team's heavily modified Humvee, had three—two on the front corners near the headlights and one in the center of the vehicle just below the radar, which itself was just below another lidar, a Riegl airborne line scanner to spot obstacles from as far as 250 feet away. Whittaker designed a gimbal to buffer the Riegl and an accompanying stereo camera and aim in the direction Sandstorm would soon be headed. The gimbal's carbon fiber shell and specially coated optical glass evoked an astronaut's helmet.[3] Carrying the load for the actual navigation were a GPS unit and an inertial measurement unit.

Whittaker's Red Team was the favorite, for good reason. It had been twenty years since Terregator took to the lawns, sidewalks, and stairways of the Carnegie Mellon campus. They were presumably furthest along on the hardest part of all this, which was creating software capable of making enough sense of all these sensors' inputs fast enough and intelligently enough not to end up in a ditch. Plus, the student-heavy team had, as one chronicler put it, "fanaticism more typical of an underdog, enviable material resources (it had become one of corporate America's most assiduous shakedown artists), and a culture of discipline that emanated from the big man himself."[4]

And indeed, Whittaker's team won, in a way: by the time Sandstorm's tires spun wildly for nearly seven minutes on a rock in a berm it got stuck on, melting the rubber off, it had gone a winding 7.4 miles and up switchback-etched Daggett Ridge with no human intervention. Had it made it over that ridge, it was smooth sailing for the next 135 miles. The entry in second, an Israeli dune buggy, got stuck in an embankment at mile 6.7. This was several times farther than TerraMax could boast, as the thirty-two-thousand-pound six-wheeler had been flummoxed by a couple of bushes at mile 1.3. The motorcycle, built by UC Berkeley PhD student Anthony Levandowski's team, tipped over at the start, its sleep-deprived creator having forgotten to flip a switch to put it into autonomous mode.[5]

For Whittaker, the "win" was pale consolation. "Mile seven? Where's that get you in life?" he tells me, years later. "That's one I gotta live with a long time."

What he has to live with is that ten days before the race, they pushed Sandstorm a bit too hard on a long training run and flipped it, mashing a quarter million dollars in hardware. They bandaged it back together, but its aim was off, and it was taking out fence posts even before it got stuck. In retrospect, the modern self-driving industry can thank Whittaker for that tumble, because had he not rolled Sandstorm, David Hall might never have invented the seminal sensor for self-driving cars.

Or maybe Hall himself would have won. His 2003 Toyota Tundra was running third when the tires of Whittaker's creation caught fire. It was only when the organizers paused the action for Sandstorm that the Tundra got hung up on its own rock and, as Hall tells me, his inability to fool the Toyota's electronic throttle prevented a concerted thrust to freedom. The path that led him to the desert traced back through the microprocessors of his battle bots, the drivers of his subwoofers, and the ripple effect of globalization.

Hall grew up in Bloomfield, Connecticut, the son of an engineer who built nuclear power plants and the grandson of a physicist who invented a scanning process to make color photographs and married into the Eastman Kodak family.

"I learned to read schematics before I learned to read English," Hall says.[6]

His grandfather helped him set up a workshop, in which, among other things, he built a guitar amp and a motorized bicycle. Even before the mechanical engineering degree at Case Western Reserve University, Hall gravitated toward projects combining the electrical and the mechanical. He set up a shop in Boston after graduation in 1974, designing electromechanical hardware for the defense, medical device, and other industries. Along the way, in 1978, he patented a handheld tachometer that counted a motor's revolutions by just touching it with the end of a shaft on the device. He licensed it to a company that called Hall's invention the Jones Computak, and royalties from it would sustain him as he ramped up his next venture.

Hall was a gifted engineer, and the idea that his creative output was being buried in the products of others began to chafe. His brother-in-law had put together a business plan related to the consumer audio market; Hall absorbed it, borrowed $250,000 from his grandfather; and, in 1983, moved to Silicon Valley to launch Velodyne Acoustics, the name in homage to his hobby of competitive cycling.

Velodyne's niche was in high-end subwoofers, ones that used a patented combination of accelerometers and a servo-controlled drive unit to erase the distortion that happens with most speakers when the input signal spikes with the volume already up. The subwoofers found devotees among audiophiles, and the Morgan Hill, California, factory employed dozens of people to build and sell them around the world.

Hall kept innovating: as early as 1990, he had developed a digital version of his subwoofer's brains using Texas Instruments digital signal processors—specialized microchips made to convert the analog signals of everyday life into the digital domain and vice versa. The industry wasn't ready. The experience came in handy, though, when as a sidelight he and his brother Bruce, who ran the customer-facing side of things at Velodyne, got into robotic pugilism at the turn of the millennium. These rumbles of remote-control robots were televised by shows such as *Robot Wars*, *BattleBots*, and *Robotica*. Hall entered the fray with metallic beasts called Drillzilla and Da Claw. This served the dual purposes of stealth marketing Velodyne subwoofers and keeping Hall's mind sharp: he had learned early that he performed best when working on more than one thing at a time, the solutions to roadblocks relating to one revealing themselves after focusing on another.

It was quickly clear that marketing to a mob that sought mechanized catharsis wasn't going to solve Velodyne Acoustics' deeper problem, which was that regardless of how good his speakers were, competitors in China were killing the business. His suppliers—the ones making speaker cones, baskets, magnets, and such—went first. Velodyne could move its production overseas, and ultimately it did. But that left Hall with a mostly empty factory in Morgan Hill, and he was looking for something to build in it. It had to be high margin and high value. The military was talking about autonomy, and they had money to spend. He would build high-end sensors for autonomous vehicles, he decided. When DARPA announced the Grand Challenge in early 2003, he saw an opening.[7]

By then, Velodyne had built and sold subwoofers controlled by digital signal processors, which Hall branded as Digital Drive. Team DAD, for Digital Auto Drive, would harness the same Texas Instruments digital signal processors Hall used for Da Claw and his latest subwoofers. Rather than servo-controlling a speaker cone or a bot with a giant drill sticking out its backside, these processors pushed the brake pedal and turned the steering wheel of his Toyota Tundra. But the magic was on top of the cab. It wasn't a lidar.

Hall used two broadcast-quality digital camera sensors to develop a stereo imager that could, he said, locate obstacles as far as 875 feet ahead. This was the same idea Hans Moravec had had for the Stanford Cart, except Hall's could support speeds of one hundred miles per hour, at least on a dry lake bed, he estimated. He used Texas Instruments digital signal processors for the stereo camera, building the circuit boards and coding them himself.[8] He didn't win the 2004 DARPA Grand Challenge, but nobody won, and he had impressed a lot of people at what he called "Woodstock for geeks" and is now recognized as the birthplace of the self-driving car industry. One of those he impressed was Jim McBride, a Ford guy who had volunteered to help DARPA keep the race safe for humans.

McBride knew lidar from his days as a laboratory condensed-matter physicist, when he worked in a Ford lab with a half dozen people dedicated to fundamental optics. He earned his PhD from the University of Michigan in Ann Arbor along the way. At Ford, McBride had used lasers to unravel the surface chemistry of the crystals in the catalytic converters that clean up car emissions and later worked with Bell Labs scientists to use a twist on atmospheric lidar to see the hydrocarbons in car exhaust. A reorg at the turn of the millennium landed him in Ford's safety department. In addition to helping out at the DARPA Grand Challenge, he was on the lookout for innovations that might one day help make the automaker's products safer.

McBride, an unconventional auto industry engineer, found a kindred spirit in Hall, an unconventional engineer, period. During the week between the qualifier and the final, the two debated how to approach autonomous sensing. Hall talked about improvements he could make to stereo vision. McBride was convinced that lidar was the way to go. Their recollections of these conversations differ in the details.

"I didn't know what a lidar was going into the first race," Hall says. "So Jim told me, you know, you send out a laser beam and you measure the time of flight, and I go, 'Huh? What?' And he said this laser works day and night. You send out the laser beam and it bounces off something and it's totally idiot-proof. So I said, well, that's nice, and put in the back of my mind as something I might look at someday. And then he said, you know, use more than one laser, 'cause one can't scan fast enough."

In McBride's recollection, "Dave had some ideas about improving his stereo vision, and I challenged him that the easiest way to get a direct perception of a 3D world around you—and this is a direct quote—I told Dave I would literally glue a hundred lasers together and spin 'em around."[9]

Hall stuck with stereo vision either way. DARPA announced a second Grand Challenge for October 2005, a year and half after the first. Hall was among 195 teams that responded to the initial call. DARPA staff did site visits to whittle down the field. They showed up at Velodyne on a rainy day in May.[10] Rainy days were among the many challenges Hall had run into when developing the stereo vision system, such as mirages, aliasing (the "wagon-wheel" effect, where something in motion looks static because of the frame rate of the viewer), and varying reflectivity. Hall by then recognized that he could address them all and get a stereo vision system to work, but that it looked like "ten years of slogging through code—through code that's a pain in the ass. And it's like, you'll get there. But I wasn't looking forward to it. I needed lower-hanging fruit than that."

Bruce sat in the driver's seat; David was in the passenger seat with a notebook computer on his lap. Despite the drizzle, the Tundra managed to avoid an orange traffic barrel in a straightaway in Velodyne's parking lot. But when the DARPA visitors set up a barrel immediately after a left turn, the Tundra plowed right into it, as Hall expected it would. The stereo vision system was forward-focused, essentially blinkered as far as peripheral vision, just like the SICK and Riegl lidar units on competitors' cars. That's why Whittaker's Riegl had a gimbal, and why his Carnegie Mellon team had used multiple SICKs to cover a wider space around the vehicle. Hall recognized that he would need three of his stereo vision systems to reasonably see around corners—and that he was burned out on stereo vision.

So as a side project to the side project of the Grand Challenge, he had already begun working on a lidar based on the seeds planted in his conversations with Jim McBride back in the Mojave desert. He started with a kit: an infrared diode laser, a laser driver, a detector driver with an amplifier. He reverse engineered the printed circuit board and the detector, looking for where the parts came from, and, in the case of the detector, ordering a smaller version to fit in what he had in mind. His idea was to put lasers and detectors all around the outside of a cylinder, like a laser-based version of a Denning ring of sonar sensors. The lasers would, combined, fire a million times per second. But unlike a static Denning ring, the entire thing would spin, up to twenty times a second. That would be his lidar's effective refresh rate as it delivered a 360-degree panorama up to 250 feet from the vehicle, covering about 28 degrees vertically. That refresh rate would be far faster than the quarter of a second or so it takes a human to react to anything.[11]

There was inspiration involved. But as with all his engineering work, perspiration dwarfed it.

"It's not an epiphany. It's a laborious mental exercise that I go through evaluating the different things, and if there's something I don't understand, I do a little experiment to convince myself that I'm right," Hall says. "But by the time I cut metal, I'm absolutely convinced I know what I'm doing and it's gonna work."

**David Hall's spinning lidar atop Team DAD's 2005 DARPA Grand Challenge Toyota Tundra. (Photo courtesy of David Hall / Velodyne Lidar.)**

Hall figured out how to deliver power as well as extract huge amounts of data from the lasers and detectors despite their spinning like a top, as well as how to establish the exact rotational position. He spent three days aligning and gluing sixty-four lasers and an equal number of detectors, each detector looking for the light reflected from a particular laser.[12] His thumb and forefinger, in protest, went numb for a month. He glued the lasers and detectors in eight clusters around a two-foot-diameter metal wheel on a somewhat larger metal box whose combined fifty pounds displaced the stereo vision system on the Tundra's roof. It looked a bit like a sombrero.

Hall enlisted help from his Velodyne engineers to get it done faster, but this was his creation. He finished the hardware two weeks before the Grand Challenge's qualifier back at the California Speedway in Fontana starting in late September 2005, with the 131.2-mile final race scheduled for October 8. He was still writing the code to train the Toyota to act on the lidar's advice during the week between the qualifier, which whittled the field down from forty to twenty-three, and the race itself.

Red Whittaker's team was back with a repaired and improved version of Sandstorm plus a new entry, a 1999 Hummer Sport Utility Truck they called H1ghlander. Both carried eleven Intel Pentium computers. One was dedicated to the Riegl lidar, another to the Riegl's gimbal, another to the four SICK lidars on each vehicle. The computer software running the Carnegie Mellon vehicles had grown to a million lines of code. Again, Whittaker was the favorite.

Jim McBride was back, this time with his own team, called Intelligent Vehicle Safety Technologies I, in a Ford F-250 Super Duty truck. While the only official references to Ford were McBride's contact email address and the listing of Ford, with Honeywell and Delphi, as a sponsor, this was the Ford team, with Littleton, Colorado–based robotics startup PercepTek as a subcontractor. McBride had seen enough in the 2004 event, he says, that the idea of full autonomy "wasn't such a crackpot idea," and that beyond enabling better safety on the roads, it could open the doors to all sorts of mobility solutions. The F-250 carried two Riegl lidars like Whittaker's, in addition to two SICK lidars and three radars.

With two entries, Whittaker decided to employ both tortoise and hare strategies: Sandstorm, the tortoise, would go slow and steady; H1ghlander would jackrabbit ahead. And indeed, H1ghlander went out fast and was far ahead when the engine started losing power when it needed it most. Slowly, the second-place vehicle caught up and, late in the race, passed the Carnegie Mellon leader. This was Stanley, a Lexus SUV carrying five SICK lidars and software written by a team led by Sebastian Thrun. Thrun, who had done mapping algorithms for Groundhog back at Carnegie Mellon, had moved on to lead Stanford's SAIL lab. Thrun took home the $2 million prize. Stanley finished in just under seven hours; Sandstorm came in nine minutes later and H1ghlander ten minutes after that. In all, five of the twenty-three teams finished, though TerraMax, the big Oshkosh military truck, did so over the ten-hour time limit.

The combination of the Ford F-250's fifty-three-foot turning radius and the vehicle's decision to avoid an encroaching military photographer at a hairpin turn conspired to send McBride's entry off the road. Hall's lidar had a mechanical problem at mile 26, one precipitated by the extra pounding suffered because of a gyroscope glitch that sent his car zigzagging four feet to the right every time it hit a bump, which often meant off the side of the road. Hall found out about it some time later, when he tried to reuse the gyroscope to make an autopilot for his boat.

"I called up the manufacturer, and he says, 'You know, that model, every one of them did that, and if we knew you were gonna use it in the Grand Challenge, we would have given you a replacement,'" Hall says.

Hall had been looking forward to another Grand Challenge, and DARPA had already announced a third desert race, perhaps assuming the second one would turn out like the first. With five finishers in the second race, DARPA canceled the third. Hall flew to Washington, DC, to protest, telling them, "You've got a whole army of people chomping at the bit for the right cause, and you just pulled the rug out from everybody." Two DARPA guys flew to California to check out Hall's lidar. They recognized the implications, and they weren't the only ones. Sanjiv Singh, who had created Cyclone, a spinning lidar with sixty-three fewer lasers, and was now a faculty member with Whittaker's Field Robotics Center, says Hall's invention enabled a new way to answer those two essential questions of autonomous motion: Where am I? and What's nearby?

Hall's lidar helped answer them in two key ways. One was an aggregate laser pulse rate that Singh calls "unheard-of," yielding an enormous amount of data. The other was the fact that it saw a twenty-eight-degree-thick ribbon of reality, everywhere at once. SICK and Riegl line scanners on the other Grand Challenge cars not only required vehicle motion or mechanical rotating or nodding to turn a line scan into a plane, but their field of view was limited to maybe sixty degrees.

"Previously we always had these lasers that pointed forward, and we had to remember what we just saw," Singh says. With, say, three sixty-degree scanners up front, you could see around the corner before making the turn. But once you made the turn and drove on, "we'd have to remember that there's a corner here." They had to construct what Singh called a "state-estimation scheme." Before Hall's invention, "keeping state," as it's called, involved combining what the sensors were seeing in real time with the "memories" of what the sensors had already seen—memories that took a lot of programming and opened the door for all sorts of complications. This had to happen over and over as the car moved along.

"That was the old way, right?" Singh says. "With the Velodyne, my god, there was so much data all the time. You could always see everything around you"—cars you had just passed, cars creeping up to pass you, bicycle couriers darting through stopped traffic, joggers popping out from blind corners, whatever. It was, from the robotic perspective, like having eyes in the back of your head.

"That's what I told David Hall. I said, 'What you did, you may not have realized. You enabled us to use a completely different set of algorithms.' Because—storing what we've seen over time—that was hard to get right, and we didn't have to do that anymore," Singh says.

DARPA announced a third race, the DARPA Urban Challenge, for 2007. Teams would get DARPA money to buy lidar units from Hall, putting Hall in the lidar business. He compressed the lidar on the Tundra into a thirty-pound stainless steel unit roughly the size of a human head, curving down elegantly from a slightly thicker top. He grouped the sixty-four lasers around two detector clusters on one side this time, but performance-wise, it was the same as the one on his Tundra. He sold the first batch of twenty for about $75,000 each, which he says was about what it cost to make them at the time.

The HDL-64, as it was called, rode with five of the six teams that completed the fifty-five-mile DARPA Urban Challenge course at a shuttered air force base near Victoryville, California. The venue was designed to test the autonomous vehicles' ability to deal with intersections, blocked roads, parking lots, lane markings, stop signs and traffic lights, moving traffic, traffic barrels, lane merges, and more. The winner was a Chevy Tahoe named Boss from Tartan Racing, as Whittaker's Carnegie Mellon team called itself this time around. In addition to the Velodyne, it carried lidars from SICK, IBEO, and Continental as well as radar and vision systems (the latter from the Israeli company MobilEye).[13] The win earned them a statue of an eagle and a $2 million check. Thrun's Stanford Racing Team, running a VW Passat called Junior, came in second this time around. In addition to the Velodyne, it carried two IBEO lidars and a camera that stitched together a panorama with six sensors.[14]

**The world according to a Velodyne HDL-64E lidar unit. (Photo courtesy of Velodyne Lidar.)**

For Hall, the Grand Challenge couldn't have come at a better time. The 2008 financial crisis sent Velodyne's slumping subwoofer business into freefall. Business selling lidars to the likes of Caterpillar (for big autonomous mine crawlers) and Navteq (for mobile mapping), kept the company going. Google's street-mapping and autonomous car efforts, led by Thrun and stocked with Grand Challenge veterans, were big early customers starting in 2008. In 2010, Hall introduced a smaller lidar specifically for mapping, this with thirty-two lasers and called the HDL-32. McBride bought them for Ford's self-driving test cars, as did many others. In 2014, Hall rolled out a sixteen-laser version and called it what it looked like: the Puck. At $8,000, it was cheaper, and automakers started mounting multiples on their autonomous research vehicles.

To recap, a subwoofer engineer racing a Toyota Tundra with a stereo vision system on its roof at "Woodstock for geeks" in 2004 ended up inventing the most important sensor in the history of autonomous vehicle development for the follow-up race 550 days later. What if Whittaker hadn't flipped Sandstorm and its sensors hadn't been skewed, and he had won that first Grand Challenge? Hall might have never retreated to his Alameda workshop and come up with his crazy sombrero of a lidar. Someone else would have eventually invented something like it. But how many years would it have taken? Where would autonomous vehicle research be today?

There's another aspect to the Grand Challenges that can't be ignored. As we've seen, government investment was crucial to all sorts of lidar development in atmospheric science, bathymetry, mapping, exploration, and defense. In this case, in addition to sparking David Hall's interest and seed funding what's today called Velodyne Lidar, DARPA's events created a center of gravity for gifted, ambitious, like-minded engineers who believed—and ultimately proved—that self-driving cars were not rolling nuclear fusion reactors that were perpetually thirty years away.

# CHAPTER 16
# THE ROAD AHEAD

I am in Pittsburgh to Uber to Uber for a ride in a driverless Uber. William "Bill" Skinner picks me up in his black Jeep Cherokee. He graduated from Princeton in the early 1970s with the football team's career receptions record. After a few years working for big food companies, Skinner spent the rest of his working career at a steel company that was called Allegheny Ludlum when he started and ATI when he saw the writing on the wall thirty years later. He did quality control—making sure the furnaces were right, the product was right, the cuts were right. He had been retired a couple of years and was bouncing around on his phone when he saw an ad for Uber drivers. He told his wife, "You know something, I'm going to find out what this is all about." He drives on Mondays, mainly, earlier rather than later when, as he puts it, "people are inebriated." He might golf with a buddy later today, if the weather holds up.

Bill Skinner is both a unique and rather typical Uber driver. I ask him for his take on self-driving cars, which could erase his retirement side gig.

"It's so far off, it's not going to be a concern for me," he says. He seems to reconsider this even as he says it. "Buts then, who would have ever thought in 1965 that you'd take a train at the airport with no one driving it?" The turn signal clicks metronomically. "It's a thing that'll come. It'll be like everything else: There will be jobs supporting the technology. You can't ever totally remove the human factor—unless you get into the *Terminator* scenarios. It'll be interesting to see what takes place. The rate of change is so fast. I'll see more in the next year than my grandfather saw in thirty years."

Skinner drops me off at the Uber Advanced Technologies Group, founded in 2015 and quickly stocked with about fifty Carnegie Mellon Robotics Institute staff. A couple of Volvo XC90s sit in the parking lot, their Velodyne

HDL-64s—pretty much the same as the ones David Hall made for the DARPA Urban Challenge a decade ago—spinning away up top. The Volvos are slate gray, matching the building and the railroad bridge and the Allegheny River and the sky. This is Uber's town under Uber's skies. Everything über-Uber.

In the lobby, the backdrop to the security desk is three stories high, a wall of big steel tiles, one of them with the word UBER cut out of it. The space is monolithic, imposing, confidence-oozing, steely-cool. I sign in, signing an NDA with the tip of my finger on a tablet. An Uber spokesperson says the company can't go into much depth about lidar and its role, given the ongoing Uber-Waymo lawsuit (Waymo being the spinoff that was Google's self-driving car project). Not a problem, I tell her—the lawsuit will be a footnote in the history of autonomous vehicle development however it turns out.[1]

We step back outside into a light rain. Madeline Auer, twenty-five, and Paul Galon, thirty, both vehicle operators, are ready to go. I slide into the black leather backseat. Galon holds his hands at the ready on either side below the wheel, which turns itself. In the passenger seat, Auer has an open notebook computer. A tablet computer mounted for the backseat passenger's benefit shows the raw feed from the Velodyne on the roof. Auer's screen fuses data from the lidar, seven cameras, and ten radar units.

"We're like the professional backseat driver," Galon says. "We nitpick everything the car does."

This car and about two hundred like it have covered a million miles in the past year or so in San Francisco, metropolitan Phoenix, Toronto, San Francisco, and Pittsburgh. A year ago, Galon says, a trip around the block was a big deal; now, the SUVs mostly drive themselves through a city that's hard for humans to navigate—three rivers making ninety-degree corners a rarity, hills everywhere, four seasons, bridges, tunnels, traffic circles, five-street intersections, construction everywhere. A quiet doorbell sound rings when the car isn't sure what to do. Galon pops a red button on the center console between him and Auer and takes over; he lifts and twists the button to put the Volvo back under its own control. On average, Uber vehicle operators have to take over maybe once every thirteen miles, which is both impressive and laggard—Waymo's vehicles apparently go 5,600 miles without interruption.[2]

The braking is silky smooth; the car corners, with what I assume is geometrical perfection. It feels like we're on rails, I say aloud. "That's how I describe it," Auer says. "Because it works on a mapped road and it knows where its lane is all the time. It knows its direction of travel. We've got the stop signs and traffic lights, but it's basically on a virtual railroad track." The virtual track is reroutable on the fly, she continues, "so if there's a car parked a little bit in our lane but not too far, we're still able to nudge around that, which is really cool."

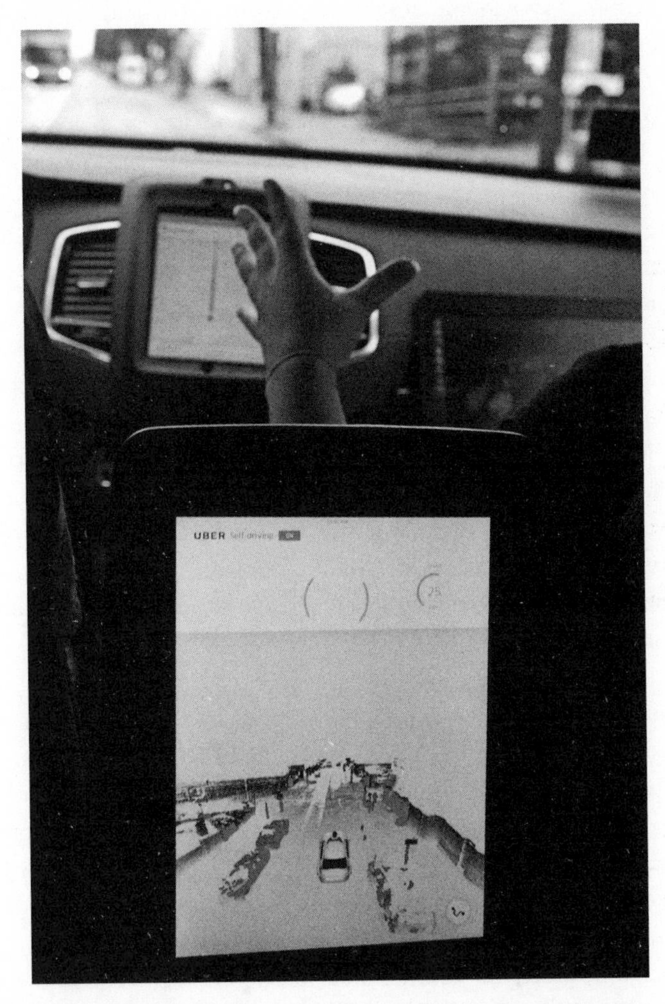


**Uber autonomous vehicle operator Madeline Auer monitors a notebook computer as a self-driving Volvo plies the streets of Pittsburgh. The screen in the foreground displays the raw data from the rooftop lidar.**

Her screen puts red envelopes around cars, yellow ones around pedestrians, and orange ones around bicycles. Galon takes over when the Volvo stops for a box truck parked perpendicular to us with its tail pushing just a bit into our lane. He does the same at construction with a flagger standing with his sign flipped to SLOW. Traffic cones infringe in our lane; a yellow Caterpillar excavator looms close. Construction is particularly confusing to the autonomous Ubers, Auer says.

"Is there a flagger? Does he have a flag up? Does he have a stop sign or a slow sign?" Auer asks. "Are there cones telling me I need to go to the other side of the road? Are those for me? Are those for somebody else? As just a human, you're like, 'I don't even know what you want me to do.'"

The self-driving Uber cars excel at dealing with pedestrians, Galon and Auer agree. Some time back, he was working at night and the car stopped hard for no obvious reason. "And we saw there was a pedestrian—it was all dark, he was dressed in black," Galon says.

"And that's a very normal thing, especially at night because the people come out of nowhere," Auer adds. "That's one thing that's awesome about lidar, that it'll pick up just everything and anything. It picks up that object, and then our car's classifying that object as a pedestrian and has a trajectory going in front of us. That's why we're here—it's for safety. And that's such a cool and amazing thing."

As we pull back into the Uber lot, I suspect that everything I write about this experience will seem quaint in a few years, as if it were about a ride in a horseless carriage or a heavier-than-air flying machine.

Someone, at some point, was going to be killed by a self-driving car. It happened at 10:00 p.m. on March 18, 2018, in Tempe, south of Phoenix, a city as easy to drive in as Pittsburgh is hard. Elaine Herzberg, forty-nine, walked her bike across three lanes from a median far from a crosswalk, apparently without taking a good look; in the fourth lane, an Uber self-driving Volvo like the one I had ridden in a few months before, unwavering as a downhill freight train, hit her. It was going forty in a forty-five zone.

Uber suspended its autonomous vehicle testing on public roads in Arizona, Pittsburgh, and elsewhere. The governor of Arizona forbade them from public roads. Officials from Waymo said their system would have done better; officials from Velodyne said their lidar most certainly saw Herzberg coming. But the future of self-driving cars suddenly looked less certain.

And there are uncertainties. But the biggest barriers may not be technical, says Victoria Waters, collaboration manager at the University of Michigan's Mcity, an autonomous vehicle proving ground that opened in 2015 and counts about every big name in the automotive business as a partner or sponsor.

We talk during a tour of sixteen acres of streetscapes designed to challenge autonomous vehicles. She drives—it's lunch hour, during which Mcity shoos away the autonomous vehicles its sponsors bring in for testing in order to mow and do maintenance. Nearly everything is sloped, by design; there are various turn lanes, bike lanes, a blind curve, a simulated tree canopy to block GPS signals, a tunnel, a railroad crossing, a stretch of dirt road, intentionally marred street signs to make it hard for cameras, a traffic circle, and a little simulated downtown area with a two-dimensional Zingerman's Deli façade among generic neighbors.

Mcity's sponsors include dozens of automakers, suppliers, and tech companies. I ask her what sorts of things they're working on. No idea, she says: "The companies are definitely very confidential about what they're doing."

But her sense is, based on the work Mcity's sponsors are farming out to the university's researchers, that the big open questions may not be technical at all, as the support of law firms and insurance companies such as State Farm and Progressive among Mcity's sponsors underscores.

"When I'm talking to people, they say, yeah, the technology is pretty mature, but a lot of other things are lagging behind, whether it's from the policy side or insurance side or consumer acceptance," Waters says.

The smart money is on computers superseding human drivers. Unlike coming up with new auto liability frameworks for self-driving cars, driving itself is not a creative act. It's rule based, a game like chess or go, if way faster-moving and a lot more complicated. It took a long time for a computer to beat the best humans at chess and even longer to beat them at go. But they did, and they will drive better than the best human drivers one day. And the best human drivers—the Formula 1 racers and the NASCAR folks who can thread through a series of croquet wickets at 175 miles an hour—they're not exactly the baseline.

Ninety-four percent of crashes are human caused (as opposed to, say, mechanical).[3] Crashes on US roads killed 37,461 in 2016, up 5.6 percent from 2015. That included 3,450 attributed to distracted drivers, 803 to drowsy drivers, and a whopping 10,497 to drunk drivers. The crashes killed 5,987 pedestrians and 840 bicyclists, the highest numbers since 1990–1991.[4] About 4.4 million people were hurt. The total cost, including deaths, injuries, medical expenses, property damage, administrative costs, and lost wages and productivity, was an estimated $432 billion that year.[5]

That's just in the United States. Globally, road traffic kills more than 1.2 million people a year.[6] Self-driving cars and trucks may never be perfect, but they are unlikely to drink and drive, text and drive, fiddle with the radio and drive, or turn to threaten fighting children and drive. Galon, the autonomous Uber operator, worked as a paramedic for years and dealt with the mess.

"I'm excited about the safety of it, given how many people I've seen hit by cars and all the accidents I've responded to," he says. "I cannot wait until it's there."

There's also the small matter of economics. Our personal automobiles—expensive, depreciating assets—sit parked 95 percent of the time.[7] Reimagining transportation as a service rather than a product, one delivered by fleets of autonomous vehicles available on a per-use or subscription basis, opens up opportunities all over the economy. Uber's motivation is simple. Remember when Netflix's core business was mailing DVDs? That's Uber with human drivers. Computers deliver a lot of video these days. Soon enough, computers will be delivering a lot of people.

Consumers are certainly ready for transportation as a service, as the explosive growth of Uber and Lyft, Gett (in Europe), Didi (China), Yandex (Russia), Ola (India), and others has shown. The Rocky Mountain Institute estimates that new mobility services powered by autonomy atop electric drivetrains (which are much more efficient and require less maintenance) could reduce annual transportation costs by $1 trillion and cut carbon dioxide emissions by a billion tons in the United States alone. Among those sharing the wealth will be the auto industry, financial institutions, governments, electrical utilities, and new entrants, some competing for business in markets that don't exist yet.[8]

The disruption could run deep. Yes, there could be a new class of supercommuters who sleep, work, or surf the mobile web in their own autonomous vehicles during suddenly less-miserable commutes. Legions of displaced truck, taxi, and ride-share drivers would need to find other work. Traffic in places like Manhattan could get worse as the convenience of door-to-door autonomous service draws people from buses and subways, as is apparently already happening with Uber and Lyft.[9] But with autonomous ride-sharing services whisking people around all day (or recharging outside of urban centers), there may also be much less need for the paved wastelands otherwise known as parking lots. One consulting firm estimates we could end up freeing up 2,200 square miles in the United States—nearly the equivalent of the area covered by all of New York, Los Angeles, Chicago, Houston, and Phoenix combined.[10]

Automakers are on board. Fiat Chrysler is building Chrysler Pacifica minivans for Waymo. GM is building autonomous Bolt electric vehicles for its own self-driving car program, which it kick-started with the $561 million acquisition of Cruise Automation in 2016. GM is working with Lyft and also intends to roll out its own autonomous ride-sharing service in big US cities in 2019 as part of a future of "zero crashes, zero emissions and zero congestion."

"We think this represents one of the biggest business opportunities of all time, since the creation of the internet," Dan Ammann, GM's president and chief of autonomous vehicle strategy, told analysts.[11]

GM's far from alone: BMW, Honda, Hyundai-Kia, Mercedes, Nissan-Renault, PSA Peugeot Citroen, Tesla, Toyota, and Volkswagen, not to mention automakers in India and China, also have serious autonomous vehicle programs. That doesn't mention a raft of suppliers and would-be suppliers such as Apple. Ford is right there too: the company intends to have a fleet-ready self-driving car on the road by 2021. Jim McBride, the optical physicist turned safety engineer turned Grand Challenge competitor, became the technical lead for autonomous vehicles for Ford's Research and Advanced Engineering group. He tells me automakers have little choice in the matter, likening the emergence of autonomous transportation to that of digital photography.

Kodak researchers invented the first digital camera in 1975, but company executives thought digital images could never displace their chemical- and paper-based mainstays.[12]

"And now, if you're not in digital imaging, you're out of business," McBride says. "And think about the immense possibilities of saving lives and giving mobility to the elderly and people who can't afford cars and reducing urban congestion and all of the things that are discussed in this space. If you're a company that's a provider of transportation in a decade or two, if you don't have a fully autonomous solution, you're going to go the road of Kodak."

Ford is spending well over a billion dollars to avoid that road, buying machine-learning/artificial intelligence outfits Nirenberg Neuroscience, SAIPS, and Argo.ai (Argo, which Ford spent $1 billion on alone, was founded by two Carnegie Mellon grads), and street-mapping startup Civil Maps.

With the exception of Tesla, whose chief, Elon Musk, has called lidar a "crutch," all these automakers consider lidar indispensable to the sensor fusion required to achieve full autonomy. What's been called the "lidar stampede" has attracted dozens of entrants. Among them: MicroVision (which uses microelectromechanical mirrors, or MEMS), Valeo (partnering with Volkswagen), Blackmore (investors include BMW i Ventures and Toyota AI ventures), Cepton Technologies, Continental (which bought Advanced Scientific Concepts' flash lidar unit), Draper, Strobe (acquired by GM's Cruise in October 2017), Ouster, Luminar (its twenty-two-year-old founder landed $36 million in funding and signed a deal to put its lidars on Toyota's self-driving cars); Innoviz (in which Canadian auto supplier Magna invested in 2017), Infineon (having bought Dutch lidar-sensor company Innoluce), Princeton Lightwave (which Ford subsidiary Argo.ai bought for its MIT Lincoln Labs–born GeigerCruizer flash lidar in October 2017), and Quanergy (which uses optical phased array technology born with Paul McManamon's defense programs and has raised $150 million from Samsung, Daimler, Delphi, and Sensata Technologies).

It is dizzying; it is ongoing. The only ones guaranteed to become major auto suppliers are the ones that are already major auto suppliers. But some will thrive, if as part of a bigger fish after the inevitable wave of weeding out and consolidation. Let's not forget, either, about Waymo, which has "spent tens of millions of dollars and tens of thousands of hours of engineering time to custom-build the most advanced and cost-effective LiDAR sensors in the industry," as the company put it in its initial complaint in the case filed against Uber.[13]

I've omitted the market leader—the only automotive lidar manufacturer capable of anything approaching mass-production volumes as of mid-2018. It's David Hall's company, Velodyne.

Velodyne Labs, as Hall's brain trust of PhD physicists, computer scientists, and optical and electrical engineers is called, moved here the month before my visit in September 2017. It occupies the entire fourth floor of a building on the north side of the island city of Alameda. Hall has had a workshop/research center in Alameda for years, and he lived on a house on a floating platform. But recently, he moved himself into a house and the research outfit into this fourth-floor office space. The space is largely empty still, the team of fifty here now expected to grow as fast as they can hire. All told, Velodyne has about 650 employees now, with just a handful keeping an eye on the legacy subwoofer business.

I'm shown a test version of Velodyne's latest creation, the yet-to-be-announced VLS-128. It has 128 lasers enabling ten times the resolution of the HDL-64 on driverless Ubers despite being one-third the size. Marta Thoma Hall sits down with me first as David Hall wrestles with some technical detail over in his small, glass-walled corner office. She's blonde with big blue eyes, stylish, and enthusiastic to the point of passionate. She's Velodyne Lidar's president and also Hall's wife, having gotten married as the subwoofer business tanked and the lidar business was just getting going. He had been divorced for a couple of years when they met online. She was and is an artist, known for big, elaborate sculptures in public spaces. He courted her in part by assisting on the engineering side, helping hang such creations as the three-story *Journey of a Bottle* suspended in the Walnut Creek Library. I ask her what he's like.

"He's always thinking. He's always dreaming. He always has his latest pet project," she says. He's perfectly fine being social, but will tell her, too, "I've been oversocialized. I need my alone time."

Hall walks in. He's as blonde as Marta, tall and solidly built, wearing aged khakis and a wrinkled oxford tucked in somewhat unevenly, a handsome man not terribly interested in maintaining appearances. In 2016, Ford invested $75 million in Velodyne, which at the same time got that same amount from Chinese search engine giant Baidu. With the strokes of those pens, Hall became a paper billionaire.

The topic quickly turns to the curved shape of the HDL-64. It was going to be a cylinder, he says. But he came up with the curves to impress his then-girlfriend.

"I think it's elegant. Beautiful! Like a sculpture," Marta says.

"Hey, a lot of work went into that." Hall is leaning way back in his chair.

"It looks like a human head, kind of. It's got nice curves. Beautiful curves."

"If I had to make it again, I'd make it a cylinder," Hall says.

"He said it was too difficult. A pain in the ass," she tells me.

"It was a pain, actually," Hall says.

With an entire automotive lidar industry taking aim at Velodyne, Hall has no shortage of pain sources. Velodyne's new "Megafactory" in San Jose, which is to automate the finger-numbing laser installation and alignment process—with the goal of being able to produce a million VLS-128s, Pucks, and other sensors in 2018—has been slow getting up to speed. Considering that the goal was once to fill an empty subwoofer factory that Velodyne's lidar production has now vastly outgrown, it's a good problem to have. But Hall has, as he puts it, "spent my entire life running manufacturing operations, and to me, there's no challenge left in it anymore. It's just poundin' out a lot of stuff, because other people run factories and they've made a lot of product in the world, so I don't get as much satisfaction out of it as I would making something novel."

He spent a thousand hours of his own time on the VLS-128, he estimates. Velodyne has also announced a solid-state product called the Velarray, which could be sold for perhaps $250 once production ramps up. As with the products of many competitors, it would take at least six units to cover the full 360-degree sweep of Velodyne's existing products. I ask him about the threat of some new entrant with an amazing new lidar technology produced on the cheap.

"This is optics that have been beaten to death since the seventies," Hall says. "Did we overlook something? Well, it's possible we did, but are we in a position to—if somebody discovers a market, can we get in there and compete with them too? I believe we can. We're flexible enough, we're not so big that we're stuck in our ways, and, you know, we sell sensors."

That last bit is a dig at the competition whose sales pitches are well ahead of their production output. Later in the conversation, Marta opens a laptop computer to show a video of one of Hall's big side projects. It shows a sleek craft with a helicopter perched on a pad on the back. Rather than a monohull, it has pontoons like missiles, notionally twelve feet in diameter. They move with the waves, but the deck is as steady as the head of a bird on a branch. Hall came up with the idea as a way to address seasickness and make boats safer and less exhausting for passengers and crew. "It's gorgeous," I blurt out.

"Never believe anything you see in a video," he says. "This is how our competitors sell lidar."

The superyacht is still in the video stage, but he's created a company called Velodyne Marine and built a less sleek, smaller version called Martini to prove the concept over the past several years, "agonizing over control systems, hydraulics, structure," he says. Among others, Google cofounder Larry Page has been out on it. Hall says he did up some slides for Elon Musk, who may or may not have been toying with the idea of installing his SpaceX Merlin rocket engines in each pontoon as "a statement piece," as Hall puts it, though the combined four million horsepower might qualify as slightly overpowered even by Musk's standards. Musk apparently remarked that he wouldn't be able to take a day off to go out on it in the next four years.

"And I sez, you know, if you don't have a day off in four years, you're probably working too hard," Hall says.

We talk about the future of the automotive lidar market and where it's going. Humans and hawks don't have laser beams firing from their eyes, after all, and they can still drive and catch rodents fairly well. Houseflies and mosquitoes avoid obstacles and judge distance with astounding skill despite compound eyes and sub-pinhead brains. Their algorithms have the advantage of having been honed over millions of years of evolutionary history. That said, there's no reason to believe that human machine vision won't catch up somewhat more quickly.

On the lidar side, Hall is convinced there's room for all sorts of sensors—spinning ones, arrays of ones with no moving parts, expensive ones, cheap ones. He wants Velodyne to make them all. He says Musk could be right about lidar as a crutch. But it's also true that lidar has been on autonomous vehicles from the moment Whittaker's team bolted the ERIM on Navlab in 1984, that it takes less computer processing (and less power) than 3D vision processing, that automotive-quality cameras "aren't free," and that "there's never been any precedent for a sensor that went on a car to ever come off a car."

"Because now you get in a car accident and the attorney says, 'Well, if you had a lidar, that wouldn't have happened,'" Hall says. "Who the hell is going to make the decision to take the lidar off the car? It's kind of there, it's gonna get designed in, and it's gonna be there for quite a while."

Better sensors will evolve, perhaps at Velodyne. But even if lidar goes the way of hand-crank starters, it will have left its mark. Lidar was and is central to building the brains that run self-driving cars—heavily informing the artificial intelligence that machine learning's virtual experiences reinforce. So regardless of what happens with Velodyne, Hall's contributions will have been epochal.

He's not unaware of this. Part of the reason he abandoned the machine shop in Boston back in 1979 was a desire for his work to be recognized. He told Marta just the other day, "If no one admired it, what's the point?"

"Well, OK," Hall says. "But that's on a customer-by-customer basis. And they get these things and they get a smile on their face. That's what it's all about." He pauses. "What people do with it, you know, I don't care if people take it and throw it in the ocean. I just enjoy making the stuff. And I enjoy showing off that I'm clever. So to me, it's just showing off what I can do that nobody else can do. I earned it in my life." He smiles just a bit mischievously. "You know, it ain't braggin' if you can back it up."

# CHAPTER 17

# GLUE, DRONES, AND RADIOACTIVE PIPES

It's not often someone founds a company, leaves, spends twenty-six years elsewhere, and then comes back. But that's what Jim West did. West was into machine vision early, starting as a General Motors Institute co-op student in 1973, where he got involved with GM's first machine vision project. A GM researcher, Bob Dewar, was running it. A few years later, a banker suggested West and Dewar found a company. They attracted a serial entrepreneur, Dwight Carlson, to be the CEO. Carlson was, as mentioned, great friends with Bill Brown, ERIM's president, and was intrigued with what lasers might do for manufacturing. West, Dewar, and Carlson landed $2 million in venture capital to found Perceptron in 1981. Their first project was to build a helium-neon laser to measure the exploding front of an inflating airbag—airbags being a new thing at the time. That led to a sheet metal measurement system, which led to a project that had West working with Leica to turn the ERIM lidar into a commercial product. That product landed on Carnegie Mellon's Navlab and other robots.

West left Perceptron in 1988 and worked with Simon Raab at FARO, among other places. His successors developed the Perceptron lidar into a product they called the Lasar. It had the same spinning and nodding mirrors familiar to the Perceptron and the ERIM. This one, though, was designed for the shop floor and found its biggest market in the lumber industry. Perceptron worked with forest products company USNR to combine the Lasar with software that could take the lidar-measured shape of a log, consider lumber market prices, and optimize the cuts in the lumber mill to maximize profit. USNR has since licensed the technology, which it still sells as the Lasar 2.

The bulk of Perceptron's modern business harks back to its automotive roots. Lasers are as fundamental to building cars as they are in enabling self-driving cars. West, the company's vice president of engineering since returning in 2014, walks me back to a high bay at Perceptron's Plymouth, Michigan, headquarters. Boxy red Perceptron instruments point from various angles at vehicle unibodies, doors, and other components. In factories around the world, Perceptron lasers position the robots and equipment that later assemble vehicles and their subassemblies; they align bumpers, wheels, vehicle frames, windshields, and so forth; they perform final gap and flush checks; and they guide welders and other robots. The eyes of Navlab have become the eyes of the factory floor.

Perceptron and competitors' products are to no small degree to thank for the far higher quality of cars today than in decades past. Those old enough to remember road travel in the 1980s and earlier will recall cars that leaked water, that rattled, that squeaked and inflicted wind noise so forceful that you'd have to yell. The doors and glass didn't fit right, and there were uneven gaps between different body panels or on opposite sides of the same panel.

"Once you take the variation out of a structure, all that goes away," says Carlson.

Carlson, Perceptron's founding CEO, has moved on to another startup he launched, called Coherix, in Ann Arbor a few miles west of Perceptron. This one focuses on lidar in auto manufacturing too. But Coherix is about glue, mainly.

"As the world tries to put together dissimilar materials, you don't weld, you don't rivet—you glue them," Carlson says. "They don't like to call it glue. It's 'structural adhesive' or 'room-temperature vulcanization' or 'sealing.'"

The bodies of today's cars are increasingly glued together, Carlson tells me, and it makes them stronger and more rigid. Glue has another benefit, he says: it can help designers be more creative because you don't have to worry about having to weld different materials together. Coherix uses lidar at close range to inspect the width, height, and volume of beads of glue. Too much is bad; not enough is bad. Skips and neckdowns are bad. It's a three-dimensional problem well suited to lidar, he says.

Lidar touches automotive transportation in another, very different way. In 1989, Colorado-based Laser Technology, Inc., introduced the first laser-based police speed gun. They pitched it as law enforcement's next salvo in an arms race that saw widespread use of radar detectors giving drivers advanced warning about traditional police radar guns. Their lidar had a footprint of less than two feet across at five hundred feet. A police radar's footprint varied, but it typically ranged from 80 to 130 feet across at the same distance. That was great for detecting speeds across multiple lanes but also made it easy for a radar detector to pick up.[1]

In 1990, *Popular Science* included the company's LTI 20/20 prototype in its "Best of What's New" issue, which GEICO insurance's president happened to peruse. Frustrated with radar detectors and the actuarial risk they emboldened, he had GEICO lend the startup the money to develop its prototype into a commercial product, which launched in 1991.[2] Companies with names like DragonEye Technology and Stalker Radar also now sell into the police lidar market. With their products' proliferation have come lidar detectors and jammers as the cat-and-mouse game continues.

There's another auto-focused lidar company to note—as much for the nonautomotive applications of their technology. As with Optech and Riegl, the German company Polytec was founded by a physicist, Heinz Lossau, in 1967. He didn't get into the business that would become Polytec's calling card, laser vibrometry, until the 1990s.

Lidar can harness the Doppler effect to monitor the motion of wind or incoming ballistic missiles. It can also use it to see how things subtly move back and forth. Vibrometry's applications are wildly diverse. The military uses it to classify targets—say, a truck or a tank behind camouflage or a boat off in the distance—based on how their engines rumble.[3] Spies can eavesdrop on conversations using a lidar to detect the vibrations of spoken words against windows, curtains, even paper coffee cups.[4]

Polytec's vibrometers are mostly used in structural analysis in the automotive and aerospace industries, in testing the dynamics of buildings and bridges based on how they vibrate, and in designing stereo and cell phone speakers, circuit boards, and disk drives, among other products. But its laser vibrometers have also attracted medical researchers who have used them to study bone dynamics, the pressure in blood vessels, and how hearing works, among other things.[5]

Not so similarly, lidar vibrometry has helped entomologists hear insects. Scientists have used Polytec's and others' laser vibrometers to monitor the high-pitched calls of crickets and the low-pitched mating calls of the invasive brown marmorated stink bug, neither of which are audible to us. They do it based on the laser's ability to pick up the vibration insects' conversations impart on the leaves. In the case of the stink bugs, which have multiplied prolifically since their arrival in 1998 and now attack a wide variety of US crops, the idea is to understand the mating calls and then interfere with them.[6]

Scientists from the US Department of Agriculture's Agricultural Research Service used a Polytec lidar in a study similar to the stink bug work, in this case to decode and disrupt the mating calls of glassy-winged sharpshooters. This bug spreads Pierce's Disease and is a vineyard scourge that costs the California grape industry $100 million a year.

"Each species has its own call, and they vary a lot," says USDA entomologist Rodrigo Krugner. "Some sound like a motorcycle. Some sound like a baby crying."[7]

A big plus: unlike spraying insecticides or releasing natural predators, vibrational mating disruption, as it's called, leaves beneficial insects alone.

Sanjiv Singh got to know the Perceptron lidar about as well as Jim West did. The Carnegie Mellon Field Robotics Center staff technologist who had wrestled with, and ultimately built, the Cyclone lidar for Red Whittaker moved into the PhD program. His work with autonomous earth movers got him interested in moving earth—not just driving it from one place to another in massive truck beds but robotically excavating it. This is a deceptively hard problem: the system has to not only perceive the shape of the area being dug as it changes with each scoop but also calculate the best place to dig next and to safely deposit it into whatever will cart it away. There was also the matter of texture: were you digging mud? Gravel? Clay? As every sandcastle builder knows, wet sand is different than dry sand. It takes a decade or longer for a human excavator operator to become expert, Singh learned, and the work is hazardous, dirty, strenuous, and repetitive, particularly at places like toxic and nuclear waste sites. After years working for Whittaker, he recognized a good opportunity for field robotics when he saw one.[8]

Singh bought a used Cincinnati Milacron hydraulic robot arm typically used in manufacturing. He built a ten-foot-square, three-foot-deep sandbox that would have struck envy in a preschooler. He poised the robotic arm over one side of it and the Perceptron lidar over the other. By then, the scanner had served on several Field Robotics Center robots and was outmoded. When Singh got the system working, it took three minutes and twenty seconds per shovelful, some of that time because the Perceptron had to be calibrated with each scoop. Four hundred experiments and much neural-network and algorithm development later, Singh had invented a new mathematical representation of the task of excavating, one incorporating resistive forces of various soil types and terrain topology into planning and digging—a first.[9] A follow-on project with Caterpillar incorporated Singh's work into an actual autonomous excavator, this one substituting two pinwheeling Riegl line scanners for the Perceptron. It dug soft earth as fast as a human operator could.[10]

Singh then shifted his attention from dirt to what's on top of it, building a lawn mower designed for golf courses, athletic fields, and other big fields in 2005. It incorporated SICK lidars, of the sort the DARPA Grand Challenge competitors used, for spotting what Singh calls "discontinuities"—rocks, shrubs, trees, garden gnomes, whatever—even if the mower was pitching up and down. His focus then shifted again, into the air.

For the US Army's Future Combat Systems program—an ambitious modernization program to incorporate all sorts of autonomy—Singh's team built a system to fly a helicopter low and avoid obstacles. Using a Fibertek lidar hung from the belly of a Honda RMax industrial remote-control helicopter, the system guided itself and avoided obstacles including trees, wires, and buildings in a demo at Fort Benning, Georgia, in 2006—also a first. But Singh's guidance system was to be part of a new autonomous helicopter, and that was scrapped along with most of the rest of Future Combat Systems' billions of dollars in hardware when the program was canceled three weeks after the Fort Benning flight. Other Singh-group contracts evaporated for unrelated reasons at the same time. Academic research costs money, and Singh's group was suddenly destitute. A small army project looking at autonomous medevac helicopters helped them scrape by for a time, but it wasn't enough to sustain the team. They went into a state of hibernation, one aided by the birth of Singh's second daughter. While too young to chip in directly, she opened the door for Singh to take paternity leave and keep the team afloat.[11]

"I was in the office every day, but the university benefits pool was paying my salary—that's how broke we were," Singh says.

He ended up back on the ground for the better part of two years, building obstacle detection systems for rolling robots designed to manage and harvest orchards, estimate grape yields with machine vision, and prune peach trees, among other things.

During this time, Singh was introduced to John Piasecki, CEO of Piasecki Aircraft, who saw the promise in autonomous flying. They teamed up to write a proposal for a federal small-business grant to develop an autonomous pilot for an army medevac helicopter. The army considered their plans too ambitious for the money involved and instead gave the contract to PercepTek—the company Jim McBride had contracted to help build Ford's entry for the 2005 DARPA Grand Challenge. But then Lockheed Martin bought PercepTek, making PercepTek part of a not-small business. Piasecki appealed and got the contract after all.[12]

Singh managed the program. His Carnegie Mellon grad student Sebastian Scherer wrote the software. Singh had hired Caltech grad Lyle Chamberlain as an engineer in essentially the same role Singh had once played for Red Whittaker. He assigned Chamberlain the development of a lidar, though without the "How hard can it be?" preamble. Chamberlain attacked the problem at the Field Robotics Center as well as at his kitchen table, machining parts while still cranking out CAD drawings. Like Singh's Cyclone, this one was built around a Riegl product, in this case a Riegl VQ-180 line scanner designed for mobile mapping and shipborne surveying.

"He worked almost nonstop for three months and single-handedly came up with something we could mount on the Unmanned Little Bird," Singh says.

That's a Boeing Unmanned Little Bird helicopter, which flew over Mesa, Arizona, with the resulting system on its nose in 2010. The Little Bird could already autonomously fly between waypoints in unobstructed terrain; Singh and colleagues' system created 3D maps from up to five hundred feet away, the goal being to detect obstacles, adjust landing paths, and even abort landings on the fly.[13] A follow-up DARPA program involving a notional flying Humvee, Transformer TX, led to a second version by 2012, which then led to work for the Office of Naval Research's Autonomous Aerial Cargo Utility System, or AACUS, starting that year. Singh launched a company, Near Earth Autonomy, to support it, while remaining on faculty at Carnegie Mellon. Chamberlain and Scherer joined as cofounders with Singh's Carnegie Mellon colleague Marcel Bergerman. By 2014, the team had a helicopter flying autonomously, in the snow, and landing without an overflight to check things out in advance.

In 2015, Singh's wilderness period in agriculture paid dividends when his team scored work for the US Department of Energy's version of DARPA (the Advanced Research Projects Agency-Energy, or ARPA-E). The TERRA-

BOOST program aimed to use remote air and ground sensing to improve yields of sorghum, a biofuel crop. Singh's team flew six- and eight-rotor drones over fields in Mexico, South Carolina, and Arizona, the drones carrying miniaturized, modernized versions of the systems of the mid-2000s. Among other efforts, Near Earth Autonomy is now working on a project in which several drones can fly through different parts of a mine to map it—versions of Groundhog and Cave Crawler that hover above the yellow mud. And they've got their own helicopter, a Bell 206L, white with blue and red stripes, nicknamed Aquafresh after the similarly colored toothpaste. In 2016, it flew the 135 miles from Zanesville, Ohio, to an airstrip a few miles north of Pittsburgh with only a stored USGS digital elevation map as a guide. Aquafresh landed about ninety feet from where it was supposed to, a position error of 0.012 percent of the distance traveled, without GPS.[14] In late 2017, Aquafresh flew across 150 miles of Ohio, from Ashtabula to Zanesville, again without GPS, estimating its speed and position with a camera and a lidar that matched scans to a stored elevation map.[15]

I hoped to watch Near Earth Autonomy drones fly out at what the team calls Nardo, a private turf-strip airport up in Allegheny County farm country named after its owner, Nardo J. Berardinelli. But rain drumrolls Singh's office window in Pittsburgh's Bloomfield neighborhood, and the trees sway with the autumn gusts. Singh is a big, energetic guy with the faintest Indian accent one could sustain without its complete demise. What's left of his hair is nearly as short as his two-day beard. We talk lidar for a while, and he walks me down the hall to the fifty-person company's shop floor/flight operations/storage area. He introduces me to the manufacturing staff, which is Gina Schirra.

"She does all the fab for us," Singh says.

Schirra is assembling a blue machined-aluminum box called Boost 2.0, to be used for TERRA-BOOST. It will carry a Velodyne Puck Lite lidar to do the mapping and state-estimating. It's a 1.3-pound little brother of David Hall's DARPA Grand Challenge lidar, with sixteen lasers spinning inside a cylinder an inch wider and about three times as thick as an actual hockey puck. Unlike a hockey puck, it collects three hundred thousand laser points per second at up to a hundred meters. The device also carries a laser spectrometer capable of assessing the state of sorghum based on the light its leaves reflect. Around the corner, a similar blue unit called Peregrine is mounted below a six-rotor DJI drone whose custom travel case is the size of a washing machine. The Velodyne Puck Lite on this one rotates on its mount like a flipped coin.

Aquafresh's flight from Ohio to Nardo gets at Near Earth Autonomy's breakthrough innovation. While the Peregrine system can make use of the DJI drone's built-in GPS, it doesn't actually need it to navigate its immediate surroundings. A Carnegie Mellon PhD student of Singh's, Ji Zhang, developed software that can take the lidar point cloud coming in from the Velodyne, create a 3D rendering of its surroundings, reconcile it with information from a tiny onboard inertial measurement unit, figure out "Where am I?" and "What's nearby?" and plot its next move without GPS—all in real time. Singh says the goal for the company isn't to build aircraft, much less operate air services, but rather to become the standard in autonomous air navigation.

**Sanjiv Singh with Near Earth Autonomy's Peregrine system mounted to a drone.**

The same technology works on the ground. Zhang and Singh cofounded another company, Kaarta, that sells handheld and portable lidar scanners based on the technology. Its offices are downstairs from Near Earth Autonomy's.

I ask Singh about his journey from Cyclone to dirt to sky.

"When you're on the bleeding edge, you really can't tell if you're on the edge of insanity or, you know, if you're actually doing something important. Most people can't see it. My experience has been, every time you do something sort of like out of the ordinary, there's a huge silence that follows. People can't—It takes them a while to digest," Singh says. One example, he says, came after the DARPA Urban Challenge in 2007. Google didn't launch its pioneering self-driving car project until 2009. "For a year a two, there was nothing. People were trying to think about, what the hell is this about? What do you do with this stuff? How do we use it?"

Four days before my October 2017 visit, Boeing HorizonX, the company's venture capital arm, confirmed that they saw what the hell Singh's technology was about, announcing an undisclosed investment and a product development partnership with Near Earth Autonomy. In addition to the relatively low-hanging fruit of military logistics and cargo delivery to remote areas, they have their eyes on drone deliveries, bridge inspections, and autonomous copters ferrying people around urban areas.[16]

What used to be the old power plant is hard to recognize now, integrated as it is into Carnegie Mellon's Newell-Simon Hall. A great steel tank reminiscent of something dotting the fringes of an oil refinery occupies one end of the old Field Robotics Center high bay (there's another high bay now, over in the Gates Center for Computer Science, plus space in the Robotics Institute's main facility in Pittsburgh's Lawrenceville neighborhood). The tank is for testing autonomous robots designed to swim in nuclear fuel-rod pools. Above, there's a metal catwalk connecting higher floors of Newell-Simon on either side. Mounted on the wall next to it are photos of Sandstorm and Boss, the DARPA Grand Challenge contestants, corners curling from the backings.

From up here, you can see, through glass-latticed arches that used to separate outside from inside, down into the machine shop and welding station. On an office door, a sign reads, "Broken robot? Kick it or something. That might work." Below the catwalk, there is what might best be described as clutter: yellow straps cast aside like ribbons after Christmas, tarps, drills, an auto battery on a desk, spools of cable, a gas can, abandoned computer hardware, a couple of traffic lights in repose on a row of gray file cabinets, assorted lumber, and a jumble of white plastic piping. And there's Cave Crawler, its tires still stained yellow from its sodden mine visits thirteen years before, one wheel up on a wooden box as if poised to autonomously spelunk again.

Red Whittaker greets me, big and still powerfully built at sixty-six, in jeans and a green T-shirt from a farm near the nine-hundred-acre one he owns in Somerset County. His watch is a Timex. The first order of business is coffee, which he makes in a pot you'd see on a diner's back counter. It's in a nook the uninitiated would need a Near Earth Autonomy drone to find again. The cup he pours into proves to be far too small, so a few minutes later, he brews another pot in a kitchen space on a different floor. He has brought me here to see an old appliance, of sorts, sitting on top of a refrigerator. As the coffee percolates, he places the appliance on a table for a closer look. It is perhaps the last, and certainly the finest, of the Denning-style ultrasonic sensing rings, this one with thirty-two transmitters and as many receivers, each no bigger in diameter than a pencil. Its innards of cleanly arranged circuit boards and rainbows of wire are exposed. Teruko Yata, a Carnegie Mellon postdoc, created it sometime between her arrival in 2000 and her death in 2002. There's a memorial lecture in her name now. The most recent speaker was Chris Urmson, a Whittaker deputy on the DARPA Grand/Urban Challenges who since has led Google's self-driving car program and founded autonomous vehicle systems developer Aurora Innovations. Lidar had rendered Yata's ring a relic by the time it was built, but Whittaker respects it for what it is and its role in a continuum of robotic sensors culminating in the likes of Velodyne's. "That's nice craft, right?" he says.

The coffee is ready; he refills his little white cup about halfway, topping it off with water. He looks sleep-deprived, though this seems to be a permanent state of being for him.

"There's still people around here that sleep every night," he says.

"You don't sleep every night?" I ask.

"No," he says.

"You just go straight through?" I ask, regretting it before it's out.

"In the vernacular of the trade, it's called an 'all-nighter,'" he says, and then chuckles.

Whittaker has been pulling more than the usual all-nighters over the past few months, for a project that could save years of work and tens of millions of dollars at some of the US Department of Energy's biggest nuclear facility cleanups.

"It takes some doing, and going into it, it's not at all clear—the fact that it's possible," Whittaker says. "And those are the ones—" He pauses and changes gears. "First, it *meant* something. It could have been beyond the possible. And from time to time, they're the ones that are worth doing."

He walks me back into the high bay for a look at what he's talking about. But first, he detours to the orderly open space in the center of the high bay, where David Kohanbash, a senior research programmer working for Whittaker, is at his desk amid three stout blue robots the size of end tables. Green LEDs indicate that they're awake. He's working on software to command a swarm of hundreds of robots outfitted with arms to assist in assembly or logistics in a manufacturing environment. Most of the robots, Kohanbash explains, are virtual. "It's kind of a mixing of real-world and simulation," he says. I note no lidar on these machines. Rather, it's a sort of Denning ring of twenty cameras around each robot's midsection. "One of the lidars is, say, $1,000 or $1,500, whereas each of these cameras you can buy on Amazon for $4.50," he says.

Whittaker then walks me over into the clutter below the catwalk. He points out a white metal box with a glass window. It's on the floor amid, for lack of a better word, junk. Above the glass window is the word "Perceptron."

"I saved this from the dumpster," he says.

We both take it in for a moment. "There it is," I say.

"I buried people trying to get this thing to work," he says.

Past Cave Crawler lies a line of two ten-foot sections of black steel pipe a golden retriever could walk through. The dog would segue straight into a forty-two-inch-diameter Easy Pour paperboard concrete forming tube. A few feet away, near a wall on which hangs Hyperion, an autonomous robot with mountain bike wheels (sponsored by NASA, it roamed Chile's Mars-like Atacama Desert in 2003), is one of the objects that's been keeping Whittaker up at night. It has three arms on each end, each ending with a small wheel; a long, narrow waist has its cover open and wires exposed like on C-3PO's belly. There are several small sensors and lidars. It evokes a two-headed metallic praying mantis. Called PipeDream, it was designed to crawl through miles of pipes of varying dimensions, surveying radioactive deposits as it went.

It started at the 2017 Waste Management Symposia in Phoenix. Rob Rimando, the director of the US Department of Energy's (DOE) Office of Technology Development in the department's Office of Environmental Management, chaired a panel. Marty Reibold, who directs strategic initiatives for something called Fluor-BWXT Portsmouth, was on the panel. So was Whittaker.

**Red Whittaker and the PipeDream prototype in the Field Robotics Center's Newell-Simon Hall high bay.**

Fluor-BWXT Portsmouth is a company formed to carry out the decommissioning of the Portsmouth site, near Piketon about seventy miles south of Columbus, Ohio. This used to be the Portsmouth Gaseous Diffusion Plant. Built from 1952–1956, it and similar facilities in Oak Ridge, Tennessee, and Paducah, Kentucky, supplied the US nuclear weapons complex with weapons-grade uranium as well as fuel for nuclear subs and ships and, later, power plants. At Portsmouth's three enormous two-story buildings (their combined square footage was greater than that of the Pentagon by the size of an Amazon warehouse), hundreds of massive compressors and converters rammed millions of pounds of uranium hexafluoride gas through about a hundred miles of pipe, separating the fissile—nuclear-fission-ready—uranium-235 from the uranium-238 that makes up 99.3 percent of natural deposits.[17]

The first building, X-333, had forty-nine football fields of floor space on two floors (2.83 million square feet) and about fifteen miles of pipes with diameters ranging from three inches to four and a half feet. That building's gaseous output of 3–4 percent U-235 then piped into a second building, X-330. It had 2.4 million square feet and twenty-five miles of pipes up to about three feet in diameter. Its output could ramp up U-235 content to about 10 percent. The third big building, the 2.5-million-square-foot X-326, could concentrate the fissile uranium to greater than 90 percent.[18]

It has all been coming down since 2001, with two thousand people working on it over the past few years, at an annual cost to US taxpayers of about $325 million. The process of characterizing, cleaning up, and carting away X-326 was already well along by March 2017, when Reibold and Whittaker addressed Waste Management conference attendees in Phoenix. X-333 was the focus now, with X-330 on deck.

The pipes were one of the biggest challenges. They were encased in ducts heated above the 134 degrees at which uranium hexafluoride gas becomes a solid. They had to be isolated from moisture in the air, which reacts with uranium hexafluoride gas to form a resin-like frost of uranyl fluoride. Over decades, moist air got into pipes to varying degrees, and the frost, called holdup deposit, collected. Uranyl fluoride is radioactive, and there are strict limits on how much of those deposits can be in a given length of pipe without risking a nuclear chain reaction, a.k.a. a criticality. In more than five years of cleanup, there hadn't been one. The bigger safety risk for the crews was the hydrofluoric acid also involved. In addition to causing burns, it can penetrate into tissue, painfully interfering with basic cellular mechanisms far from the burn site.

"It's like chemical warfare," Reibold says.

And so the hundreds of workers had to wear protective suits with respirators as they painstakingly did hundreds of thousands of radiation measurements at chalk marks they made at every foot of every pipe—each measurement taking anywhere from fifteen seconds to a couple of minutes. As the pipes tend to be far overhead, they often did so up on a cherry picker thirty feet high, one lifting a gamma radiation detector high as a colleague with a tablet computer registered the reading. The reading went to an analysis group that then wrote up a report comparing however many grams they estimated to be clinging inside a pipe based on the readings to a "critical index." If it was above that theoretical point at which it could spark a criticality when the pipe got demolished with the building, workers would have to cut out the pipe and scrape out the radioactive frost.

All this opened the door to human error, Reibold says, and because they were measuring through nickel-lined steel pipe often running parallel to similar pipe, they often found that supposedly hot sections had no holdup deposits at all.

Reibold knew Whittaker as a prominent roboticist and also as the founder of RedZone Robotics, which he had spun out of Carnegie Mellon to focus on robots to do dirty work at nuclear and other hazardous or unpleasant places. After the panel, Reibold introduced himself and told Whittaker that DOE had already spent $100 million characterizing the pipes of Portsmouth and still had two gargantuan buildings to go. Reibold and colleagues had considered some sort of pipe crawler but couldn't think of a way to make it fail-safe.

The next morning, Reibold returned to the conference center. "I've got my cup of coffee, I'm still kind of sleepy," he tells me. "I haven't made it ten steps in the door, and there's Red. 'I've been thinking about your problem,' he said. It's like, holy shit, Red, let me have my coffee."

Reibold brought DOE's Rimando into the conversation. Rimando's office pays for innovations along these lines. They met each day through the end of the conference. They settled upon the idea of a robot that would become PipeDream. A spinning lidar on its nose would identify obstacles or valves in the pipe and determine the pipe's dimensions. Lidars on a spinning Plexiglas disk toward the back would measure the inside of the pipe very precisely—be it metal or holdup deposit. Inductive sensors also on the Plexiglas disc would measure where the pipe's metal was. The holdup deposit for a given section of pipe would be the difference between what the lidar measured and what the inductive sensor measured. It would, Whittaker made clear, "really push the limits of physics," but it would work.

Rimando approved the project. Reibold told Whittaker he had 120 days—they were getting going on the next building by the end of July and had to make a decision.

Whittaker needed a team, fast. "And you couldn't pull enough, even from an empire like this," he says. He got "a dirty dozen, and some of 'em were green. But a lot of those very young people could really pour in the hours and go strong and put the back-to-back time into it. And you make up a lot." We have been walking; he stops and turns to me. "A lot of times, you get that choice in a person, between experience and commitment. And I'll take commitment. Or deep skill and expertise and commitment, and I'll take the commitment."

He assembled a group including undergrads and even a high-school student. But there were also experienced hands, among them Kohanbash;

Heather Jones, a senior project scientist; and Chuck Whittaker, a senior Carnegie Mellon field robotics specialist who has worked with his brother Red since the Terregator days. Chuck had also worked at RedZone and could speak the nuclear world's language.

We're in a hallway just beyond the high bay, having drifted here from PipeDream, when Red tells me all this. He walks me to Chuck's closet-like, hardware- and paper-crammed office adjacent to a machine shop and disappears for more coffee. Chuck, shorter, stockier, and nine years younger, bears zero resemblance to his sibling. We talk for a bit; then Chuck walks me around a corner and through an indoor space that was once the loading dock upon which Navlab was constructed. We end up in an area hosting a red Jeep Wrangler: Navlab 11, with modular racks and old SICK lidars. Behind it, rare-earth magnets poached from deceased hard drives press old blue-and-yellow NAVLAB 5 and NAVLAB 8 license plates against a metal bookshelf. Above them are posters of Navlabs past (a Pontiac Bonneville, a Houston transit bus, a map showing the Navlab 5 Pontiac Trans Sport's route across the country). Chuck walks to the back of the room and, with help from John Kozar, the modern-day Navlab lead, lifts a blue metal box from the ground to the table so I can check it out. It's about the size of a microwave oven and has a horizontal window. Its bottom mounts are corroded and its many screws rusty, but there it is: the ERIM. And then we're off to a meeting.

The meeting is in a combination storeroom (big yellow crates against one wall stenciled with "PROPERTY OF FIELD ROBOTICS") and electronics testing area. Eight team members are here, standing or in old office chairs or folding chairs like the one Whittaker leans back in. The meeting is about a new robot called RadPiper. It's necessary because, despite the team's success in creating something that could measure holdup deposits in a radioactive pipe to an accuracy of 0.4 millimeters (narrower than the lead of a mechanical pencil) while moving several feet per minute, PipeDream had a fatal flaw: the uranium deposits weren't necessarily homogenous. That is, some could be more concentrated. Taylor Wilson, consulting on the project, had said the solution would be to detect radiation directly. "It takes the guesswork out of what the enrichment is, the thickness is, and the concentration of the material," Wilson tells me later. Just twenty-three, Wilson's celebrity among geeks approaches that of Whittaker: at age fourteen in 2008, Wilson built a nuclear fusion reactor in his parents' Arkansas basement; he had his own nuclear physics lab at the University of Nevada, Reno, when he was still in high school. He is a global authority on all things nuclear.

Whittaker's team switched gears, into a crash project to build something they called DiscoBot. Its essence was to put a radiation detector between two round lead-and aluminum plates held vertically so only radiation from a one-foot segment of pipe could enter. They had until late September to finish it and compare the results with those of PipeDream and with the chalk-and-cherry-picker approach.

"He was driving his team, and these kids worked like dogs—it's like slave labor," Reibold says. "But they had a blast, all hyped up on Red Bull and running experiments until three in the morning. He has just an incredible team, and they would follow him to the ends of the Earth."

DiscoBot came out on top during tests at Portsmouth in late September and early October 2017. Now, in late October, Whittaker's team sits in a meeting, embarking on a project to build an upgrade called RadPiper. It will incorporate elements of both its predecessors and have the flexibility to assess a variety of pipes' diameters. They'll need lidar for navigation and to register the geometry of fittings, reducers, and couplings, not to mention registering the pipe's shape, which affects radiation readings. Lidar could also spot troublesome lumps of uranium deposit in which fissile U-235 lurks under layers of U-238. In the meeting, Whittaker's presence is of calm control. He suggests possibilities and then lets young team members work out the details. They discuss needs such as 153 feet of thirty-inch pipe and radiation sources for testing. It's going to be another tough one, but the team seems unawed. Ryan O'Keefe, a project engineer fresh off his master's degree, walks in and shows the group something tiny in his palm.

"It's an americium source," he says. "Zero-point-nine microcuries."

"Where'd you get it?" someone asks.

"It fell out of a smoke detector," O'Keefe says.

"Don't eat it," Whittaker suggests.

Reibold thinks Whittaker's robots can work through about half of the 15.3 miles of pipe in the next building, X-333; in X-330, it could be 12.5 miles. He figures the robots can do it with eight times less labor than the status quo without factoring in human error. "It can safely save tens of millions of dollars, and there's a good chance that we could save more," he says. That doesn't include the cleanup at Portsmouth's sister site at Paducah, either.

The applications could go well beyond nuclear site cleanup. One of the biggest questions in nuclear weapons nonproliferation has to do with what's called MUF, or "material unaccounted for." When inspectors go into a plant and find that two tons of plutonium went in and only a ton and a half came out, "the classic response is 'The rest of it is MUF'—it's in the pipes somewhere," Reibold says.

"In a facility like Portsmouth, you can have hundreds of nuclear bombs' worth of material in the piping, and you want to make sure it isn't diverted to nefarious actors," Wilson adds. "If we can figure out technologies like robotic inventorying of nuclear material, it's going to really change things—even things like the Iran nuclear deal. Those kinds of agreements are really all about developing technologies and protocols to make sure it is and will remain a peaceful program."

They're shooting to have a production model of RadPiper doing real measurements at Portsmouth by mid-2018. For Whittaker, the robot may be new, but the challenge is familiar.

"On a project like this, everything counts. So if you don't do everything, you haven't done anything. And because of that, every person counts, and every action counts, even the ones that turn out to be blind holes," he says.

I ask him if this has been a universal truth across his many projects. He answers without hesitation.

"Yes."

This book must end, but new ways to use lidar continue to emerge. As discrete tools or in combination with other technologies, as is the case with RadPiper and self-driving cars, lidar's ability to measure distance, characterize matter, and create 3D digital renderings of the world around us is broadly seductive to scientists and engineers.

As lasers evolve and advance, lidar's horizons expand accordingly. With $2.1 million in ARPA-E funding, University of Colorado researchers led by Greg Rieker in March 2018 finished up an atmospheric lidar that can detect methane leaks continuously, from miles away. At least 2 percent of US natural gas production leaks away—nine million tons, costing $30 billion a year—and it's also a greenhouse gas eighty-six times as potent as carbon dioxide.[19] Rather than the single wavelength of standard lasers, Rieker's lidar uses an optical frequency comb emitting pulses of more than a hundred thousand wavelengths at once (the 2005 Nobel Prize in Physics went to the optical frequency comb's progenitors). The light bounces off mirrors a mile or more away, and if methane or other trace gases are in the air, they absorb the light in ways the lidar can detect. Rieker and colleagues say their device can spot leaks over many square miles, continuously, with a thousand times the sensitivity of existing technologies.[20]

David Harding, the Goddard Space Flight Center scientist, is working with laser physicist colleague Anthony Yu on a lidar that could incorporate several laser colors at once, including a tunable beam Yu is developing. Such a system could help scientists one day observe photosynthesis globally from space. That, in turn, could help estimate how well trees and plants are producing sugars and, ultimately, how much biomass those sugars will produce. With such a system, one could even infer foil moisture and nutrients or whether there's a bark beetle infestation in a previously healthier forest.[21]

There's room for fun too. Montreal-based multimedia entertainment house Moment Factory bought a Velodyne Puck lidar and developed software to let users play a more colorful version of the 1970s-era videogame Pong, but on a forty-by-sixty-foot virtual box painted in light from above. Players solo or in tandem move paddles projected at their feet by shuffling/hopping/dancing back and forth on their edge of the box, which the lidar picks up. Moment Factory debuted the technology at the 2017 MAPP_MTL festival, the idea being to bring gaming out into the real world and trigger spontaneous social interactions, which it did.[22]

We've seen lidar enable mapping, measuring, autonomous driving (and flying), digital preserving and discovering, insect-eavesdropping, natural-gas-leak sniffing, Pong playing, and much more. Divining where new generations of bold and creative people like those we've met will one day innovate would be as foolhardy as Hutchie predicting any of the above. But two things are certain: autonomous vehicles will soon drive lidar out of its historic anonymity, and whatever happens, lidar's future is going to be at least as interesting as its past.

# ACKNOWLEDGMENTS

Having stumbled into a topic so broad and so deep, I relied heavily on many generous folks to educate me directly and steer me in the right direction for further investigation. Some of them, such as Arlen Chase, John Degnan, Brian Edwards, Gary Guenther, Bill Krabill, Tom Painter, Sanjiv Singh, and Red Whittaker, were notable players in the narrative. Others, including Jeff Deems, David Harding, Susan Jones, Paul McManamon, Michael Raphael, Johannes Riegl Jr., Xiaoli Sun, John Weishampel, Jim Yungel, and John Zayhowski, provided me far more help and guidance than their brief mentions in this book indicate.

Others to whom I'm indebted got no mention at all: Don Neilson and Rebecca Michals of SRI, Michele Durant at HRL, Audrey Lengel and Allison Rein of the Niels Bohr Library and Archives at the American Institute of Physics, Paul LaRocque of Teledyne Optech, David Murphy of Near Earth Autonomy, Nora Kazour at Carnegie Mellon, Jeff Bennett, Don Carswell, Bob Samson, David Smullin, Ron Swonger, and Guido Visconti. And I'm sure I'm forgetting someone.

The University of Colorado Denver's Auraria Library kindly allowed me guest researcher access, helping enable the depth of reporting I was hoping for. UCHealth continued to let me tell its stories while working on this book. Jacqueline May Parkison did a masterful job editing the manuscript. I'm grateful for the support of literary agent John Willig and of the team at Prometheus Books. Finally, I thank my wife and daughters for their patience.

# NOTES

## CHAPTER 1. HUTCHIE

1. John F. Donegan, Denis Weaire, and Petros S. Florides, *Hutchie: The Life and Works of Edward Hutchinson Synge (1890–1957)* (Pöllauberg, Austria: Living Edition Science, 2012), 5.

2. Email from Denis Weaire, Sept. 28, 2016. In Yeats's 1923 Nobel Prize speech, in which the great poet mentioned J. M. Synge a dozen times, he remembered his old friend as "a strange man of genius." "He was the only man I have ever known incapable of a political thought or of a humanitarian purpose," Yeats said. "He could walk the roadside all day with some poor man without any desire to do him good, or for any reason that he liked him."

3. W. J. McCormack, *Fool of the Family* (New York: New York University Press, 2000), 128.

4. Donegan, Weaire, and Florides, *Hutchie*, 14.

5. Ibid., 4, 14.

6. McCormack, *Fool of the Family*, 395, 406.

7. Ibid., 407–8.

8. Donegan, Weaire, and Florides, *Hutchie*, 18.

9. Ibid., 19.

10. Ibid., 56.

11. McCormack, *Fool of the Family*, 407.

12. Cathleen Synge Morawetz, telephone interview, September 28, 2016. Morawetz carried on the Synge predilection for brilliance as a lauded mathematician and New York University faculty member. In 1998, she became the first female mathematician to win the National Medal of Science.

13. Donegan, Weaire, and Florides, *Hutchie*, 41.

14. Ibid.

15. John H. Lienhard, "No. 524: Einstein: Inventor," *Engines of Our Ingenuity*, accessed September 23, 2016, http://www.uh.edu/engines/epi524.htm. Einstein and coinventor Leo Szilárd, who would go on to conceive the nuclear chain

reaction, safely rank as the most overqualified duo in the annals of refrigeration-technology development.

16. Donegan, Weaire, and Florides, *Hutchie*, 44.

17. Ibid., 48.

18. E. H. Synge, "A Method of Investigating the Higher Atmosphere," *Philosophical Magazine* 7:9, no. 60 (1930): 1014.

19. It's worth noting that Hutchie also recognized the potential of instrument-bearing rockets to reach much higher, citing (among a grand total of four references) Robert Goddard's pioneering 1920 report "A Method of Reaching Extreme Altitudes," a largely theoretical piece by an American physicist some thought a crackpot. Even by 1930, when Hutchie published his searchlight paper, Goddard's experimental rockets had attained heights topping out at ninety feet.

20. Synge, "Method," 1020.

21. Ibid., 1015.

22. M. A. Tuve, E. A. Johnson, and O. R. Wulf, "A New Experimental Method for Study of the Upper Atmosphere," *Terrestrial Magnetism* 40, no. 452 (1935).

23. E. O. Hulburt, "Observations of a Searchlight Beam to an Altitude of 28 Kilometers," *Journal of the Optical Society of America* 27, no. 377 (1937).

24. "To Explore beyond the Stratosphere with Light," Smithsonian Science Service Historical Images Collection, accessed February 21, 2017, http://scienceservice.si.edu/pages/122026.htm.

25. E. A. Johnson et al., "The Measurement of Light Scattered by the Upper Atmosphere from a Search-Light Beam," *Journal of the Optical Society of America* 29, no. 512 (1939): 515–16.

26. Ibid., 516.

27. "Ellis A Johnson," INFORMS, accessed February 21, 2017, https://www.informs.org/About-INFORMS/History-and-Traditions/Biographical-Profiles/Johnson-Ellis-A. The only subsequent work of note with searchlights was done in 1950–1952 by Air Force Cambridge Research Center scientist Louis Elterman. See Louis Elterman, "Seasonal Trends of Temperature, Density, and Pressure to 67.6 km Obtained with the Searchlight Probing Technique," *Journal of Geophysical Research* 59, no. 3 (September 1954): 351–58.

28. E. H. Synge, "A Design for a Very Large Telescope," *Philosophical Magazine* 7:10, no. 63 (1930): 353.

29. Ibid., 358.

30. Donegan, Weaire, and Florides, *Hutchie*, 32.

31. McCormack, *Fool of the Family*, 476.

32. Donegan, Weaire, and Florides, *Hutchie*, 33.

33. McCormack, *Fool of the Family*, 407.

34. Donegan, Weaire, and Florides, *Hutchie*, 20.

35. Jeremy R. Cummings, Thomas J. Fellers, and Michael W. Davidson, "Introduction to Near-Field Scanning Microscopy," Olympus Microscopy Resource Center, accessed September 23, 2016, https://www.olympus-lifescience.com/en/microscope-resource/primer/techniques/nearfield/nearfieldintro/.

36. Ling Zang, "Lecture 16: Near-Field Scanning Optical Microscopy," 32, accessed September 23, 2016, http://www.eng.utah.edu/~lzang/images/Lecture_16_NSOM.pdf.

37. J. Roger, P. Angel et al., "The Multiple Mirror Telescope," in *Telescopes of the 1980s*, ed. G. Burbidge and A. Hewitt (Palo Alto, CA: Annual Reviews, 1981), 69.

38. Anthony A. Stark et al., "The Antarctic Submillimeter Telescope and Remote Observatory (AST/RO)," *Publications of the Astronomical Society of the Pacific* 113, no. 783 (2001), doi:10.1086/320281.

## CHAPTER 2. ENTER THE LASER

1. Georgi Dalakov, "Basile Bouchon," History of Computers, accessed January 29, 2018, http://history-computer.com/Dreamers/Bouchon.html.

2. Charles G. Gross, "The Fire That Comes from the Eye," *Neuroscientist* 5, no. 1 (1999): 59. For Aristotle's terminology, which is roughly as convoluted as the underlying notion, see R. D. Hicks, *De anima* (Cambridge: Cambridge University Press, 1907): II–7, 77–83.

3. David C. Lindberg, "Alhazen's Theory of Vision and Its Reception in the West," *Isis* 58, no. 3 (Autumn 1967): 322, 325. Alhazen's logic failed to fell the theory. Even in the twenty-first century, a surprising number of people seem to subscribe to it. See G. A. Winer et al., "Fundamentally Misunderstanding Visual Perception. Adults' Belief in Visual Emissions," *American Psychologist* 57, no. 6–7 (2002): 417–24.

4. Mario Bertolotti, *The History of the Laser* (Bristol, UK: Institute of Physics Publishing, 2005), 102.

5. Ibid., 106–7; "This Month in Physics History: Einstein Predicts Stimulated Emission," *APS News*, August/September 2005, http://www.aps.org/publications/apsnews/200508/history.cfm.

6. Martin Hollman, "Christian Huelsmeyer, Inventor," Radar World, accessed August 7, 2017, http://www.radarworld.org/huelsmeyer.html.

7. "This Month in Physics History - April 1935: British Patent for Radar System for Air Defense Granted to Robert Watson-Watt," *APS News*, April 2006, https://www.aps.org/publications/apsnews/200604/history.cfm.

8. Ibid.

9. "MIT Radiation Laboratory," MIT Lincoln Laboratory, accessed November 28, 2017, https://www.ll.mit.edu/about/History/RadLab.html.

10. Joan Lisa Bromberg, *The Laser in America, 1950–1970* (Cambridge, MA: MIT Press, 1991), 2, 63.

11. Ibid., 71.

12. Jeff Hecht, "The Race to Make the Laser: From Maser to Laser," author's website, accessed November 28, 2017, www.jeffhecht.com/pioneers.html. This mirrored cavity, called a Fabry-Pérot interferometer, had been around for sixty years.

13. "About Theodore Maiman," Laser Inventor: Creator of the World's First Laser, accessed November 28, 2017, http://www.laserinventor.com/bio.html.

14. Bromberg, *Laser in America*, 40, 84, 86.

15. "Ruby Crystal from Maiman Laser," Smithsonian National Museum of American History, accessed November 29, 2017, http://americanhistory.si.edu/collections/search/object/nmah_711120. This assumes Maiman's pinkie fingertip was roughly three-eighths of an inch in diameter and eleven-sixteenths of an inch long.

16. Bromberg, *Laser in America*, 87, 91.

17. Joan Bromberg, "Interview of Eric Woodbury," Niels Bohr Library & Archives, American Institute of Physics, November 14, 1985, www.aip.org/history-programs/niels-bohr-library/oral-histories/4977.

18. Bromberg, *Laser in America*, 126.

19. Bromberg, "Interview of Eric Woodbury."

20. Ibid.; *The 1962 Aerospace Year Book* (Washington, DC: American Aviation Publications, 1962), 112. The Soviets weren't far behind: by 1964, scientists at the Vavilov Optics State Institute had developed their own laser range finders. See Vasil Molebny, Paul McManamon, Ove Steinvall, Takao Kobayashi, and Weibiao Chen, "Laser Radar: Historical Prospective—from the East to the West," *Optical Engineering* 56, no. 3 (December 2016): 1.

21. David Smullin, "Memories from David Jordan: Louis Smullin," Facebook, April 11, 2010, https://www.facebook.com/groups/116048505785/permalink/10150662198305786/; Joel Moses, telephone interview, August 29, 2017.

22. David Smullin, telephone interview, September 21, 2017; Paul Penfield

Jr., "Louis D. Smullin, 1916–2009," in *Memorial Tributes* 17 (Washington, DC: National Academies Press, 2013), 308–13, https://www.nap.edu/read/18477/chapter/51#308.

23. "Louis Smullin, Former Electrical Engineering Department Head, 93," *MIT News*, June 8, 2009, http://news.mit.edu/2009/obit-smullin-0608.

24. Louis D. Smullin and Giorgio Fiocco, "Optical Echoes from the Moon," *Nature* 194 (June 30, 1962): 1267; Brian Edwards, telephone interview, November 3, 2017.

25. Louis D. Smullin, "Education for Life or Life-Long Education?," *Annals of Geophysics* 46, no. 2 (April 2003): 435; Bromberg, *Laser in America*, 155.

26. Smullin, "Education for Life," 438; Guido Visconti, "Giorgio Fiocco (1931–2012)," International Radiation Commission, accessed December 1, 2017, http://www.irc-iamas.org/files/EOS_per_Fiocco.pdf; Glauco Benedetti-Michelangeli et al., "Giorgio Fiocco: A Jolly Good Fellow and His Research," *Annals of Geophysics* 46, no. 2 (April 2003): 182.

27. Smullin, "Optical Echoes," 1267; J. W. Graham et al., "Radio Astronomy," Research Laboratory of Electronics (RLE) at the Massachusetts Institute of Technology, *Quarterly Progress Report*, no. 66 (July 15, 1962): 60, 63, https://dspace.mit.edu/handle/1721.1/53764.

28. Ibid., 61–62.

29. Smullin, "Education for Life," 435.

30. Graham et al., "Radio Astronomy," 62.

31. Smullin, "Education for Life," 435.

32. Walter Sullivan, "Man Shines a Light on the Moon," *New York Times*, May 11, 1962.

33. Sullivan, "Man Shines a Light"; James Ring, "The Laser in Astronomy," *New Scientist*, June 20, 1963, 672–73; Smullin, "Optical Echoes."

34. Sullivan, "Man Shines a Light."

35. Susan B. Jones, telephone interview, September 14, 2017.

## CHAPTER 3. INTO THIN AIR

1. Gerald W. Grams, "Laser Radar Measurements of the Aerosol Content of the Atmosphere," *NAS-NRC Atmospheric Exploration by Remote Probes* 2 (1969): 208, https://ntrs.nasa.gov/search.jsp?R=19720017710.

2. Joan Lisa Bromberg, *The Laser in America, 1950–1970* (Cambridge, MA: MIT Press, 1991), 113–14.

3. Giorgio Fiocco and Louis Smullin, "Detection of Scattering Layers in the Upper Atmosphere (60–140 km) by Optical Radar," *Nature* 199 (September 28, 1963). In regard to the laser power: it's about cramming more photons (and thus power) into smaller and smaller packages to deliver the energy pumping into the laser. The packages are defined by time. The RCA laser's packages had a pulse width of fifty nanoseconds, which was ten thousand times shorter than the half a millisecond of the Raytheon laser. So it could pack ten thousand times more photons in a pulse, assuming the same input power. Luna See's Raytheon laser had a hundred times more energy pumped into it, though, which cut the pulse-for-pulse advantage of the RCA laser to one hundred.

4. Fiocco and Smullin, "Detection of Scattering Layers."

5. Paul Ligda, "Myron George Herbert Ligda," Ligda Lineage: The Great Lineage of the Ligda Family, accessed December 4, 2017, http://www.ligda.net/myron-george-herbert-ligda/.

6. W. F. Hitschfeld, "The Invention of Radar Metrology," *Bulletin of the American Meteorological Society* 67 (January 1, 1986): 33.

7. Roger C. Whiton et al., "History of Operational Use of Weather Radar by U.S. Weather Services. Part I: The Pre-NEXRAD Era," *Weather and Forecasting* (June 1998), https://doi.org/10.1175/1520-0434(1998)013<0219:HOOUOW>2.0.CO;2; Myron G. H. Ligda, "The Laser in Metrology," *Discovery*, July 1965, 7.

8. Hitschfeld, "Invention of Radar Metrology," 34.

9. "Scientists Follow a Storm, Spot a 'Hook,' Warn a Town," *Life*, April 16, 1956, 40.

10. Ligda, "Myron George Herbert Ligda."

11. Donald L. Nielson, *A Heritage of Innovation: SRI's First Half Century* (Menlo Park, CA: SRI International, 2004), 9–23.

12. Ligda, "Myron George Herbert Ligda"; Paul Ligda, telephone interview, December 5, 2017.

13. Nielson, *Heritage of Innovation*, 9–23; Charles A. Northend, "Lidar, a Laser Radar for Meteorological Studies," *Die Naturwissenschaften* 54, no. 4 (1967): 78–79.

14. "The Rapid Development Pace of LIDAR," *SRI Journal* no. 2 (1964 [reprint]): 2.

15. "Ronald Thomas Collis," Legacy.com, accessed December 5, 2017, http://www.legacy.com/obituaries/sfgate/obituary.aspx?n=Ronald-Thomas-Collis&pid=20201866.

16. Warren Johnson, telephone interview, June 20, 2017.

17. R. T. H. Collis and John Oblanas, "Project Pre-Gondola II. Airborne

Lidar Observations," US Army Engineer Nuclear Cratering Group, January 1968, http://www.dtic.mil/dtic/tr/fulltext/u2/735657.pdf. This was part of the Plowshare program, an Atomic Energy Commission effort to see if nuclear weapons technology could be adapted for peaceful uses such as blasting giant tunnels and canals, digging immense holes, and shaking natural gas loose from tight deposits in the days before fracking. The SRI team used ground-based lidar in 1966 and flew in 1967; the explosions were, fortunately, chemical.

18. Edward E. Uthe, "Lidar Environmental Observations," *Proceedings of SPIE* 2112 (3 June 1994): 2, doi: 10.1117/12.177305; Johnson, interview.

19. Warren B. Johnson, "Lidar Observations of the Diffusion and Rise of Stack Plumes," *Journal of Applied Meteorology* 8 (June 1969). The first study of smokestack effluent looks to have been done by a team in England at the Tilbury Power Station in 1966. See P. M. Hamilton, "The Use of Lidar in Air Pollution Studies," *International Journal of Air and Water Pollution*, no. 10 (1966).

20. "Barbados Oceanographic and Meteorological Experiment (BOMEX)," NCAR Earth Observing Laboratory, accessed December 7, 2017, https://www.eol.ucar.edu/field_projects/bomex.

21. Johnson, interview; "Frequently Asked Questions, Subject A17): What Is the Saharan Air Layer? How Does It Affect Tropical Cyclones?" NOAA Hurricane Research Division, Atlantic Oceanographic & Meteorological Laboratory, accessed December 7, 2017, http://www.aoml.noaa.gov/hrd/tcfaq/A17.html.

22. Johnson, interview; Uthe, "Lidar Environmental Observations," 2; Nielson, *Heritage of Innovation*, 9–25.

23. You need two lasers to be able to subtract the scattering from other molecules and aerosols (which would affect both lasers equally) from the one tuned to whatever you're trying to measure.

24. R. T. H. Collis, "Lidar," *Applied Optics* 9, no. 8 (August 1970): 1786.

## CHAPTER 4. PLANE CRASHES, HAILSTORMS, AND DISTANT PLANETS

1. In slightly more technical terms, the laser is acting as its own local oscillator in a heterodyne detection system. Because the speed of the air mass being measured is a tiny fraction of the speed of light, the frequency of light coming back is very close to that of the light the laser just sent out. The returning light and the local-oscillator light get mixed together on the photodetector, and the difference between them manifests in a beat frequency that's equivalent to the difference

between the incoming and outgoing laser frequencies. The beat frequency is a lot slower and much easier to deal with electronically.

2. Todd Alhart, "Edison's Heir: Bob Hall's Invention Lit Up the Future," *GE Reports*, December 1, 2016.

3. Joan Lisa Bromberg, *The Laser in America, 1950–1970* (Cambridge, MA: MIT Press, 1991), 159.

4. H. Z. Cummins, N. Knable, and Y. Yeh, "Observation of Diffusion Broadening of Rayleigh Scattered Light," *Physical Review Letters* 12, no. 6 (February 10, 1964).

5. See Mike Smith, "Defeating the Downburst: 20 Years Since Last U.S. Commercial Jet Accident from Wind Shear," *Washington Post*, July 2, 2014.

6. Eric Adams, "Boeing Planes Could Fire Laser from Their Noses to Spot Turbulence," Wired.com, September 15, 2017.

7. Michael Huerta, "'NextGen Helps FedEx Deliver Valentine's Day Gifts on Time,'" Federal Aviation Administration, February 12, 2015; "Infographic: Wake Turbulence Separation Standards for Aircraft," US Department of Transportation Volpe Center, accessed December 11, 2017, https://www.volpe.dot.gov/content/infographic-wake-turbulence-separation-standards-aircraft. Also, Michigan Aerospace has developed a clear-air turbulence Doppler lidar system that uses ultraviolet light rather than the more standard infrared. The shorter wavelength bounces off not aerosols in the atmosphere, but the air molecules themselves. See http://www.michiganaerospace.com/experience/air/.

8. See Paul F. McManamon, "Review of Ladar: A Historic, yet Emerging, Sensor Technology with Rich Phenomenology," *Optical Engineering* 51, no. 6 (June 2012): 8–9.

9. Among other ways, this is done by turning a lidar into a twist on a long-distance spectrometer. Called Raman lidar, it uses a very specific laser wavelength known to jostle particular target molecules (nitrogen, say) such that rather than simply bouncing the same frequency of light back, the target molecules vibrate or rotate, so the backscattered light returns in a shifted frequency as a result of what's called Raman scattering. As long as you're here, it's worth mentioning that atmospheric scientists call Raman scattering "inelastic." That's opposed to "elastic" scattering, in which photons bounce back with the same frequency (Doppler effect notwithstanding). To further complicate matters, there are two main types of elastic scattering. Rayleigh scattering, which happens with molecules/aerosols of diameters less than about one-tenth of the wavelength of the laser's light, and Mie scattering, which happens when the molecules/aerosols have a diameter larger than about the laser light's wavelength.

10. G. Fiocco et al., "Measurement of Temperature and Aerosol-to-Molecule Ratio in the Troposphere by Optical Radar," *Nature* 229 (January 18, 1971).

11. It works like this: Air molecules move really fast (hundreds of miles per hour). Aerosols move with the wind, so a lot slower. Because of their much higher speed, the light bouncing back to the lidar's detector is generally Doppler shifted much more than the light bouncing off the aerosols moving with the wind (that is, it's "bluer" if the molecules happen to be moving toward the detector; it's "redder" if they're moving away from it; and it's the same color if the molecules are moving lateral to the detector when the photon pings it). By filtering out the returning laser light from more highly Doppler-shifted oxygen and nitrogen molecules, Eloranta's HSRL can work out how much of a signal is actually from the aerosol backscatter. See Päivi Piironen, "A High Spectral Resolution Lidar Based on an Iodine Absorption Filter," (PhD diss., University of Wisconsin, 1994), http://lidar.ssec.wisc.edu/papers/pp_thes/thes_4.htm.

12. University of Wisconsin-Madison HSRL Data Archives, accessed December 14, 2017, http://hsrl.ssec.wisc.edu/.

13. "GV-HSRL," NCAR Earth Observing Laboratory, accessed December 14, 2017, https://www.eol.ucar.edu/instruments/gv-hsrl.

14. Matt Hayman and Scott Spuler, "Demonstration of a Diode-Laser-Based High Spectral Resolution Lidar (HSRL) for Quantitative Profiling of Clouds and Aerosols," *Optics Express* 25, no. 24 (November 27, 2017).

15. A diode laser (or laser diode) is the most common modern laser, found in laser printers, supermarket scanners, Blu-Ray players, fiber-optic telecom hardware, and of course cat-taunting laser pointers. They are tiny and pumped not by flashlamps or other lasers, but rather by electrical current, which kicks off the stimulated emission of photons that bop back and forth in a tiny gap in a semiconductor and then emerge as coherent laser light.

16. Aldo Svaldi, "Damage from Last Year's Massive Front Range Hail Storm Cost $2.3 billion—$900 Million More Than First Estimate," *Denver Post*, May 7, 2018.

17. National Research Council Committee on Developing Mesoscale Meteorological Observational Capabilities to Meet Multiple Needs, *Observing Weather and Climate from the Ground Up: A Nationwide Network of Networks* (Washington, DC: National Academies Press, 2009), 31. There are other, less weighty factors that play into weather prediction, including air pressure, soil moisture, and something called hydrometeor mixing ratios.

18. "A Brief History of Upper-Air Observations," National Weather Service, accessed July 11, 2017, https://www.weather.gov/upperair/reqdahdr.

19. "Maps of the NWS Rawinsonde Network," National Weather Service, accessed July 11, 2017, https://www.weather.gov/upperair/nws_upper.

20. There are other ways of figuring out atmospheric moisture. An interesting example is GPS Integrated Precipitable Water, which looks at radio signal delays between GPS satellites and ground stations. There are about four hundred of these GPS receivers reporting hourly across the United States. Spuler's lidar would be more accurate. See National Research Council Committee, *Observing Weather,* 97.

21. Rit Carbone, email message to author, July 13, 2017.

22. Rit Carbone, email message to author, July 20, 2017; National Research Council Committee, *Observing Weather,* 83.

23. Carbone, email, July 13, 2017.

24. S. M. Spuler et al., "Field-Deployable Diode-Laser-Based Differential Absorption Lidar (DIAL) for Profiling Water Vapor," *Atmospheric Measurement Techniques* 8 (2015). Earlier (and existing) water-vapor-sensing lidar relied on big, powerful lasers and enough hardware to fill a shipping container, and it cost $10 million to $20 million.

25. Carbone, email, July 13, 2017; National Research Council Committee, *Observing Weather*, 10, 100.

26. M. R. Bowman, A. J. Gibson, and M. C. W. Sandford, "Atmospheric Sodium Measured by a Tuned Laser Radar," *Nature* 221 (February 1, 1969). The sodium layer, roughly three miles thick and between fifty-six and sixty-five miles up, is where sodium is neither bound to other molecules (below) nor ionized (above). It had been discovered forty years earlier by Lowell Observatory astronomer Vesto Slipher, who should be much more famous. Slipher also discovered galactic redshift, which Edwin Hubble used to come up with his big bang theory, and he launched the project that led to Clyde Tombaugh's 1930 discovery of Pluto.

27. Gardner, interview.

28. R. Foy and A. Labeyrie, "Feasibility of Adaptive Telescope with Laser Probe," *Astronomy & Astrophysics* 152 (January 1985): L29–L31.

29. Laird A. Thompson and Chester S. Gardner, "Experiments on Laser Guide Stars at Mauna Kea Observatory for Adaptive Imaging in Astronomy," *Nature* 328 (July 16, 1987). Thompson's web write-up describes a nice example of multidisciplinary research in action: see www.lairdthompson.net/aohist.html. MIT Lincoln Laboratory researchers independently invented the same thing in secret. See Thomas H. Jeys, "Development of a Mesospheric Sodium Laser Beacon for Atmospheric Adaptive Optics," *Lincoln Laboratory Journal* 4, no. 2 (1991).

30. This iron Boltzmann lidar used solid-state alexandrite lasers to send out two slightly different ultraviolet beams, one at 372 nanometers and the other at

374 nanometers. Iron has different resonance lines at each of those frequencies, and they calculated temperatures based on the ratio of the fluorescence signals—the subtle difference in how the iron glowed. See Xinzhao Chu and George C. Papen, "Resonance Fluorescence Lidar for Measurements of the Middle and Upper Atmosphere," in *Laser Remote Sensing*, ed. Takashi Fujii and Tetsuo Fukuchi (Boca Raton, FL: Taylor & Frances, 2005), 189.

31. Xinzhao Chu et al., "Lidar Observations of Neutral Fe Layers and Fast Gravity Waves in the Thermosphere (110–155 km) at McMurdo (77.8S, 166.7E), Antarctica," *Geophysical Research Letters* 38 (2011). The first discovery of iron layers in the thermosphere was reported here; the formation mechanisms were studied in Xinzhao Chu and Zhibin Yu, reported in "Formation Mechanisms of Neutral Fe Layers in the Thermosphere at Antarctica Studied with a Thermosphere-Ionosphere Fe/$Fe^+$ (TIFe) Model," *Journal of Geophysical Research – Space Physics* 122 (2017).

32. Gravity waves in this instance are ripples at the edge of space caused by, for example, winds raking over mountains far below. They happen when gravity tries to flatten out the disturbance caused when two fluids of different densities interact with each other. Waves on a lake or ocean are gravity waves, in fact. Not to be confused with gravitational waves of the Einsteinian physics variety.

33. It turns out Chu was delayed until November, so Harry had left by the time she and the new McMurdo students installed a new sodium Doppler lidar at the Arrival Heights Observatory in the austral summer season of 2017–2018. Since January 2018, the two lidars (iron Boltzmann and sodium Doppler) have run in parallel to unlock secrets of the SAIR, determine the fluxes of different meteoric metal species entering the Earth's atmosphere, and understand how those metal species and SAIR respond to different solar activities, geomagnetic storms, and atmospheric waves.

34. Chester S. Gardner et al., "OASIS: Exploring the Interaction of Earth's Atmosphere with Space," *Atmospheric and Geospace Sciences Community Report submitted to the National Science Foundation*, January 2014.

## CHAPTER 5. TAKE THE PLUNGE

1. George D. Hickman, telephone interview, November 10, 2016.

2. James S. Bailey, "ONR's Lidar Program," in *The Use of Lasers for Hydrographic Studies*, ed. Hongsuk H. Kim and Philip T. Ryan (Washington, DC: NASA Scientific and Technical Information Office, 1975), 6.

3. Ibid.

4. Duane Bright, "NAVOCEANO's Lidar Program," in Kim and Ryan, *Use of Lasers*, 9.

5. Gary C. Guenther, *Airborne Laser Hydrography: System Design and Performance Factors*, NOAA Professional Paper Series (Springfield, VA: National Technical Information Service, 1985), 4.

6. G. Daniel Hickman and John E. Hogg, "Application of an Airborne Pulsed Laser for Nearshore Bathymetric Measurements," *Remote Sensing of Environment* (March 1969).

7. Bright, "NAVOCEANO's Lidar Program," 9–10.

8. Mike Rankin, "Naval Air Development Center," in *Laser Hydrography User Requirements Workshop Minutes*, ed. Lowell R. Goodman (Rockville, MD: NOAA, 1975), 49–74.

9. In more technical terms, examples of bathymetric error sources include multiple scattering propagation geometry; air/water interface reflection and volume backscatter uncertainty; beam steering and geometric stretching at the air/water interface; wave correction; hardware quantization; nonlinear signal processing and detection algorithms; and my personal favorite, spurious responses. They had to grasp the various interrelated impacts involving scan angles, altitude, receiver field of view, optical bandwidth, transmitter pulse characteristics, wind speed, and water clarity. This doesn't include integrating GPS and inertial navigation units such that the precise location of a given scan is known (more on this later). For these systems to work, researchers had to solve all these problems, and it took some very smart people many years to do it. See Guenther, *Airborne Laser Hydrography*, 2.

10. James E. Bailey, "Office of Naval Research," in Goodman, *NOAA Laser Hydrography Minutes*, 37–48; George D. Hickman, "Recent Advances in the Applications of Pulsed Lasers in the Hydrosphere," in Kim and Ryan, *Use of Lasers*, 81–88.

11. Bernard Rubin, "NASA's Lidar Program," in Kim and Ryan, *Use of Lasers*, 12–24. They flew over the oil spill with a 337-nanometer nitrogen laser pulsing one hundred times per second; algae was zapped with yellow light at 590 nanometers.

12. Hongsuk H. Kim and George D. Hickman, "An Airborne Laser Fluorosensor for the Detection of Oil in Water," in Kim and Ryan, *Use of Lasers*, 197–202. Though it remained an Airborne Oceanographic Lidar research focus for more than two decades hence, fluorosensing's utility was eclipsed by that of bathymetry on the lidar side.

13. To the surprise of no one, probably. See Rubin, "NASA's Lidar Program," 15, 18.

14. Avco-Everett Research Laboratory won the contract to build the original

version of AOL; NASA researchers Frank Hoge and Bill Krabill led the subsequent development. Jim Yungel, telephone interview, September 13, 2017.

15. This is not to say the NASA Wallops team was alone in pushing ahead with airborne bathymetry. Another system was independently invented in Australia, by a team led by Michael F. Penny. Called WRELADS (for Weapons Research Establishment Laser Airborne Depth Sounder), it first flew in 1975. See M. Calder, "WRELADS—The Australian Laser Depth Sounding System," *International Hydrographic Review*, Monaco LVII (1) (January 1980): 31–54.

16. Gary C. Guenther, telephone interview, January 4, 2018.

17. Guenther, *Airborne Laser Hydrography*, 80–82; Yungel, interview; Bill Krabill, telephone interview, January 21, 2018.

18. Guenther, *Airborne Laser Hydrography*, 100–101.

19. By the late 1970s, the Australians, the Swedes, and the Soviets had their own first-generation systems. See Gary C. Guenther et al., "Meeting the Accuracy Challenge in Airborne Lidar Bathymetry," *Proceedings of the 20th EARSeL Symposium: Workshop on Lidar Remote Sensing of Land and Sea, Dresden, Germany* (June 16–17, 2000): 3; Gary C. Guenther, email message to author, March 11, 2018.

20. Allan Carswell, "Optech: The First 10 Years, 1974–1984," white paper (July 2012): 4.

21. Allan I. Carswell, "Airborne Lidar Studies of the Hydrosphere (from the Very Beginning)" (Presentation, Teledyne Optech, 2015).

22. Ibid.

23. The infrared bounces off the surface; the green penetrates. The combination makes it easier to determine depth.

24. A.k.a. Guenther, *Airborne Laser Hydrography*. Guenther added to his corpus with a great roundup of airborne laser bathymetry through the mid-2000s: see Gary C. Guenther, "Airborne Lidar Bathymetry," in *Digital Elevation Model Technologies and Applications: The DEM Users Manual*, 2nd ed., ed. D. F. Maune (Arlington, VA: ASPRS Publications, 2007).

25. HALS was partially reincarnated, though. The NASA Wallops AOL team fetched it and used parts of it—in particular, its frequency-doubled 532-nanometer neodymium yttrium-aluminum-garnet (Nd:YAG) laser—used on later flights. See William B. Krabill and Robert N. Swift, "Airborne Lidar Experiments at the Savannah River Plant," NASA Technical Memorandum 4007 (1987).

26. Matthew Chambers and Mindy Liu, "Maritime Trade and Transportation by the Numbers," US Bureau of Transportation Statistics, accessed December 30, 2017, https://www.rita.dot.gov/bts/sites/rita.dot.gov.bts/files/publications/by_the_numbers/maritime_trade_and_transportation/index.html.

27. Jeff Lillycrop, telephone interview, November 29, 2017.

28. Ibid.; Paul LaRocque, interview, October 25, 2017; Carswell, "Airborne Lidar Studies." SHOALS stands for Scanning Hydrographic Operational Airborne Laser Survey. That first system, SHOALS-200, pulsed at 200 hertz, with the Nd:YAG laser's 1064-nanometer beam bouncing off the surface and a beam frequency-doubled to 532 nanometers for penetrating the water. Surface effects blinded the system for the first two meters or so, and its maximum depth was about forty meters in crystal clear water.

29. CZMIL stands for Coastal Zone Mapping and Imaging Lidar. In addition to the lidar, it has a passive thirty-six-band hyperspectral imager (capturing reflected sunlight) that can help determine the makeup of foliage or surface materials as well as what's in the water column and on the bottom (seagrass, mud, sand) based on the spectral signature of the reflected light. CZMIL can get to depths of eighty meters/250 feet in clear water while still being eye safe.

30. Lillycrop, interview.

## CHAPTER 6. DISCO BALL IN SPACE

1. James Faller, telephone interview, December 28, 2017.

2. Ibid.; James Faller, "Lunar Laser Ranging," NASA, 19th International Workshop on Laser Ranging, October 27, 2014, http://cddis.gsfc.nasa.gov/lw19/docs/2014/Papers/3127_Faller_paper.pdf.

3. Dicke, trained as a nuclear physicist, worked in radar at MIT's Radiation Laboratory during World War II. He transitioned to atomic physics and then got interested in gravity—in particular, in testing how well Einstein's general relativity actually described it and whether his own Brans-Dicke theory described it better. Among the ways to do that would be to track the motions of the moon or Mercury precisely and for long periods of time.

4. W. F. Hoffman, R. Krotkov, and R. H. Dicke, "Precision Optical Tracking of Artificial Satellites," *IRE Transactions on Military Electronics* MIL-4, no. 1 (January 1960). The ideas embodied in the paper, submitted in October 1959, had emerged from one of these evening physics salons; the authors noted the contributions of the other participants, including Faller and Carroll Alley, whom we'll talk about, as being coconspirators in dreaming it up.

5. Henry Plotkin, "Genesis of Laser Satellite Tracking at NASA/Goddard," NASA, 19th International Workshop on Laser Ranging, October 27, 2014, https://cddis.nasa.gov/lw19/docs/2014/Presentations/3128_Plotkin_Presentation.pdf.

6. Ibid.; H. H. Plotkin et al., "Reflection of Ruby Laser Radiation from Explorer XXII," *Proceedings of the IEEE* (March 1965).

7. Plotkin, "Genesis of Laser Satellite Tracking."

8. An entire scientific field is dedicated to this: geodesy, managed by international organizations such as the International Earth Rotation and Reference Systems Service.

9. John J. Degnan, "Thirty Years of Satellite Laser Ranging," *Proceedings of the Ninth International Workshop on Laser Ranging Instrumentation*, Canberra, Australia, November 7–11, 1994, https://ilrs.cddis.eosdis.nasa.gov/docs/ThirtyYearsOfSatelliteLaserRanging.pdf. See also John J. Degnan, "A Celebration of Fifty Years of Satellite Laser Ranging," NASA, 19th International Workshop on Laser Ranging, October 27, 2014, https://cddis.nasa.gov/lw19/docs/2014/Presentations/Degnan_Colloquium_presentation.pdf.

10. "Now 40, NASA's LAGEOS Set the Bar for Studies of Earth," *NASA History*, May 4, 2016, https://www.nasa.gov/feature/goddard/2016/now-40-nasas-lageos-set-the-bar-for-studies-of-earth.

11. John Degnan, interview, October 26, 2017; Carroll O. Alley, interview by Joan Bromberg, Niels Bohr Library & Archives, American Institute of Physics, May 19, 2006, www.aip.org/history-programs/niels-bohr-library/oral-histories/30249; C. O. Alley, "Introduction to Some Fundamental Concepts of General Relativity and to Their Required Use in Some Modern Timekeeping Systems," *Proceedings of the Precise Time And Time Interval Systems and Applications Meeting* (1981): 708–15.

12. We'll talk more about GPS, but one of the ways it works is by sending atomic time as kept on the satellites. The combination of the satellite's speed (which slows time down a bit) and lessened gravity (which speeds time up a bit more) equates to the satellite clocks running about thirty-eight thousand nanoseconds a day faster than the ones on the ground.

13. Faller, "Lunar Laser Ranging," 7. LAGEOS, developed years later, also went with 1.5-inch cubes, as have many other satellites.

14. See T. W. Murphy, "Lunar Laser Ranging: The Millimeter Challenge," *Reports on Progress in Physics* 76 (2013); and T. W. Murphy, "APOLLO: Millimeter Lunar Ranging," *Classical and Quantum Gravity* 29 (2012). The McDonald Observatory, the French Côte d'Azur Observatory, the Matera Laser Ranging Observatory in Italy, and the Apache Point Lunar Laser-ranging Operation (APOLLO) in New Mexico still take regular measurements against the various lunar retroreflector arrays.

15. Degnan's work was behind this improvement. A version of the laser used

for Alley's gravity experiment was installed in NASA's STAndard LASer (STALAS) SLR rig in 1975. In the early 1980s, a more powerful ultrashort pulse laser, built for Degnan by the French company Quantel, was installed in MOBLAS-4 through MOBLAS-8. Those MOBLASes are, as of this writing, still operating, though not mobile, in California, Australia, South Africa, and Tahiti in addition to the GGAO. John Degnan, email message to author, January 9, 2018; Degnan, email message to author, March 6, 2018.

16. MOBLAS-7, for example, fires hundred-millijoule pulses at five to ten hertz; SLR2000 would fire pulses of less than one millijoule, but two thousand times a second.

17. For more on NGSLR, see Jan McGarry, "NGLR System Overview," January 2014, https://ntrs.nasa.gov/search.jsp?R=20160011969.

## CHAPTER 7. ZAPPING MARS, MOON, MERCURY, AND EVEN EARTH

1. S. C. Cohen et al., "The Geoscience Laser Altimetry/Ranging System (GLARS)," NASA Goddard Space Flight Center, September 1, 1986, https://ntrs.nasa.gov/search.jsp?R=19870005254.

2. W. M. Kaula et al., "Analysis and Interpretation of Lunar Laser Altimetry," in *Proceedings of the Third Lunar Science Conference*, ed. E. A. King, vol. 3 (Cambridge, MA: MIT Press, 1972).

3. See "GPS Time Series," on the NASA JPL website, which shows the motion of some two thousand stations around the world over a number of years, https://sideshow.jpl.nasa.gov/post/series.html.

4. H. Jay Zwally, telephone interview, September 15, 2017.

5. L. C. Rossi et al., "Mars Observer Radar Altimeter Radiometer," NASA Godard Space Flight Center, January 1, 1988, https://ntrs.nasa.gov/search.jsp?R=19890018749.

6. Jim Abshire, interview, October 27, 2017.

7. Ibid.

8. Xiaoli Sun, interview, October 26, 2017. We're talking Nd:YAG lasers at 1064 nanometers.

9. Ibid; Eric Desfonds, email message to author, November 7, 2017. Desfonds is the manager of Excelitas Canada's aerospace sensors product line.

10. Greg Neumann, interview, October 27, 2017; M. T. Zuber et al., "The Mars Observer Laser Altimeter Investigation, *Journal of Geophysical Research* 97, no. E5 (May 25, 1992).

11. See "BMD Laser Radar," Forecast International report, October 1996, https://www.forecastinternational.com/archive/disp_old_pdf.cfm?ARC_ID=464; Thomas B. Coughlin et al., "Strategic Defense Initiative," *Johns Hopkins APL Technical Digest* 13, no. 1 (1992).

12. "LITE . . . A View from Space," NASA Langley Research Center, https://www-lite.larc.nasa.gov/.

13. See "The Shape of 433 Eros from the NEAR-Shoemaker Laser Rangefinder," accessed January 15, 2018, http://sebago.mit.edu/near/nlr.30day.html. Maria T. Zuber, an MIT scientist, led the NEAR-Shoemaker Laser Rangefinder science team. She was also a key player on the MOLA, MLA, and LOLA instruments.

14. This came in the form of a waveform digitizer that could sample the returning pulses in up to one hundred intervals that were as short as two nanoseconds. See Jack Bufton et al., "Shuttle Laser Altimeter: A Pathfinder for Space-Based Laser Altimetry & Lidar," NASA Goddard Space Flight Center, 1996, https://ntrs.nasa.gov/search.jsp?R=19960003752; and Jack L. Bufton, David J. Harding, and James B. Garvin, "Shuttle Laser Altimeter: Mission Results & Pathfinder Accomplishments," NASA Goddard Space Flight Center, 1999, https://ntrs.nasa.gov/search.jsp?R=19990087525.

15. Sun, interview; Xiaoli Sun, email message to author, March 11, 2018.

16. Sun, interview; Xiaoli Sun, "Light Detection and Ranging (LIDAR) from Space—Laser Altimeters," in *Wiley Encyclopedia of Electrical and Electronics Engineering*, ed. John Webster, article published August 15, 2016, https://doi.org/10.1002/047134608X.W8320. One of the more interesting results of MLA, as the Mercury Laser Altimeter was known, came from material in the deep polar craters of Mercury. Their bottoms were so black that no laser light returned. It turns out that this was because sooty material, left over from the thuds of comets, absorbed it all.

17. "NASA Beams Mona Lisa to Lunar Reconnaissance Orbiter at the Moon," *NASA News*, January 17, 2003, https://www.nasa.gov/mission_pages/LRO/news/mona-lisa.html; Xiaoli Sun et al., "Space Lidar Developed at the NASA Goddard Space Flight Center—The First 20 Years," *IEEE Journal of Selected Topics in Applied Earth Observations and Remote Sensing* 6, no. 3 (June 2013), doi:10.1109/JSTARS.2013.2259578.

18. "Cloud Aerosol Transport System (CATS)," NASA Goddard Space Flight Center, accessed March 5, 2018, https://cats.gsfc.nasa.gov/.

19. Space Studies Board Committee on the Decadal Survey for Earth Science and Applications from Space, *Thriving on Our Changing Planet: A Decadal Strategy for Earth Observation from Space* (Washington, DC: National Academies Press, 2018), 3–51 (see table 3.5, "Observing System Priorities").

20. In the early 1990s, NASA considered putting a wind-sensing lidar on a big satellite called the Earth Observing System. Called LAWS, for Laser Atmospheric Wind Sounder, it would have involved orbiting a big carbon dioxide gas laser. See Lockheed Missiles & Space Company, "Phase II Design Definition of the Laser Atmospheric Wind Sounder (LAWS)," vol. 2: Final Report (November 1992).

21. "MERLIN: A Satellite to Measure Methane," CNES, accessed March 5, 2018, https://merlin.cnes.fr/en/MERLIN/index.htm.

22. NASA Global Climate Change, "Graphic: The Relentless Rise of Carbon Dioxide," accessed January 19, 2018, https://climate.nasa.gov/climate_resources/24/.

23. Renee Cho, "Why Thawing Permafrost Matters," *Phys.org*, January 12, 2018, https://phys.org/news/2018-01-permafrost.html.

## CHAPTER 8. LAND AND ICE

1. Waleed Abdalati, interview, September 7, 2017.

2. W. B. Krabill et al., "Airborne Laser Topographic Mapping Results from Initial Joint NASA/U.S. Army Corps of Engineers Experiments," NASA Technical Memorandum 73287 (June 1980), https://ntrs.nasa.gov/search.jsp?R=19800019281.

3. Jim Yungel, telephone interview, September 13, 2017.

4. Think of a carrier wave as the wave upon which the information you're sending surfs. With a car radio, a station broadcasting at 99.5 FM does so at a frequency of 99,500,000 hertz, which is about five thousand times higher than a human ear can hear. But that's just the carrier wave's frequency. Radio stations take care of business by encoding the carrier wave to deliver the far lower frequencies prescribed by Bach or played by Bachman-Turner Overdrive.

5. Bill Krabill, telephone interview, January 21, 2018.

6. William B. Krabill and Chreston F. Martin, "Aircraft Positioning Using Global Positioning System Carrier Phase Data," *Journal of the Institute of Navigation* 34, no. 1 (Spring 1987).

7. Krabill, interview.

8. Craig S. Lingle and Douglas R. MacAyeal, "Robert H. Thomas, 1937–2015," International Glaciological Society, February 26, 2015, https://www.igsoc.org/news/bobthomas/. Ice shelves have water below them; ice sheets have land below them. Melting ice shelves don't raise sea level because they're already in the sea. Melting ice sheets add ice to the sea like fresh cubes in a glass of water.

9. J. H. Mercer, "West Antarctic Ice Sheet and $CO_2$ Greenhouse Effect: A Threat of Disaster," *Nature* 271 (January 26, 1978).

10. Ibid.

11. Robert H. Thomas, Timothy J. O. Sanderson, and Keith E. Rose, "Effect of Climatic Warming on the West Antarctic Ice Sheet," *Nature* 277 (February 1, 1979); Carolyn Gramling, "Just a Nudge Could Collapse West Antarctic Ice Sheet, Raise Sea Levels 3 Meters," *Science*, November 2, 2015, http://www.sciencemag.org/news/2015/11/just-nudge-could-collapse-west-antarctic-ice-sheet-raise-sea-levels-3-meters; Jeff Tollefson, "Giant Crack in Antarctic Ice Shelf Spotlights Advances in Glaciology," *Nature News*, February 20, 2017, https://www.nature.com/news/giant-crack-in-antarctic-ice-shelf-spotlights-advances-in-glaciology-1.21507.

12. The first radar altimeter in space was on Skylab in 1973, accurate to about one-hundred-meter vertical resolution. So GEOS-3 was quite an improvement.

13. R. L. Brooks et al., "Ice Sheet Topography by Satellite Altimetry," *Nature* 274 (August 10, 1978).

14. H. Jay Zwally et al., "Surface Elevation Contours of Greenland and Antarctic Ice Sheets," *Journal of Geophysical Research* 88 (January 1, 1983). The coverage here too was limited by Seasat's orbit, which maxed out at seventy-two degrees latitude north and south.

15. Radar altimetry has come a long way since the 1980s. The European Space Agency's CryoSat-2 spacecraft, optimized for sea ice and ice sheets and launched in 2010, manages to get vertical resolution of 1.3 centimeters (about half an inch) across 250-meter-diameter spots.

16. H. J. Zwally, R. H. Thomas, and R. A. Bindschadler, "Ice-Sheet Dynamics by Satellite Laser Altimetry," *Proceedings of the IEEE International Geoscience and Remote Sensing Symposium*, Washington, DC, June 8–10, 1981.

17. An inertial navigation system (INS), a.k.a. inertial measurement unit (IMU), uses gyroscopes, accelerometers, and other sensors to track the orientation and velocity of the aircraft.

18. They also got creative in automating the connection between the onboard GPS receivers and the P-3's autopilot system. Wayne Wright, an engineer on Krabill's team, hacked the instrument landing system typically used when landing in poor visibility. He tricked the system into steering the aircraft through ILS tones from antenna on the aircraft itself. ILS works by broadcasting slightly different tones from the left and right side of a runway, which are in balance if the aircraft is on the right trajectory. In this case, the tones balanced when, according to GPS readings, the aircraft was flying over previous tracks. Jim Yungel, telephone interview, September 13, 2017.

19. W. Krabill et al., "Greenland Ice Sheet: High-Elevation Balance and Peripheral Thinning," *Science* 289 (July 21, 2000).

20. Abdalati, interview. It's worth noting a key point he made about how glaciologists keep track of glaciers in the modern age. One is radar interferometry, which can measure how fast glaciers are flowing and whether they're speeding up or slowing down. A second way is essentially weighing the ice sheets using the GRACE gravity-sensing satellites, which gives bulk mass and, over time, how that mass changes. Lidar is the third, which gives elevation and shape change. Without lidar, he says, "we wouldn't have the insights as to how the shape of the ice sheet is changing, and that's important for understanding how the ice is changing."

21. H. Jay Zwally, telephone interview, September 15, 2017.

22. GLAS's Nd:YAG lasers operated at 75 millijoules at 1064 nanometers and at 35 millijoules at 532 nanometers. Both shot six-nanosecond pulses at forty hertz.

23. H. Jay Zwally et al., "Mass Gains of the Antarctic Ice Sheet Exceed Losses," *Journal of Glaciology* 61, no. 230 (2015).

24. H. Jay Zwally et al., "Greenland Ice Sheet Mass Balance: Distribution of Increased Mass Loss with Climate Warming; 2003–07 versus 1992–2002," *Journal of Glaciology* 57, no. 201 (2011). About two-thirds of the annual three-millimeter sea level rise in recent years is from the melting of ice sheets and glaciers; most of the rest is from thermal expansion of water taking up more space as it warms.

25. R. Kwok et al., "Thinning and Volume Loss of the Arctic Ocean Sea Ice Cover: 2003–2008," *Journal of Geophysical Research* 114 (July 7, 2009); Sinead L. Farrell et al., "Five Years of Arctic Sea Ice Freeboard Measurements from the Ice, Cloud and Land Elevation Satellite," *Journal of Geophysical Research* 114 (April 23, 2009).

26. See the NOAA Arctic Program's annual Arctic Report Card for the latest information about Arctic warming. Available at http://arctic.noaa.gov/Report-Card.

27. An airborne lidar called SIMPL (Slope Imaging Multi-polarization Photon-counting Lidar), which had both photon-counting green 532-nanometer and infrared 1064-nanometer beams, helped ease concerns that collecting data with a different laser would make the results hard to square with those of the first ICESat. David Harding, telephone interview, September 15, 2017.

28. "Technical Specs," ICESat-2 mission website, accessed January 23, 2018, https://icesat-2.gsfc.nasa.gov/science/specs; Nicholas W. Sawruk et al., "ICESat-2 Laser Technology Readiness Level Evolution," *Proceedings of SPIE* 9342, no. 93420L (February 20, 2015), doi: 10.1117/12.2080531; "NASA Assessments of Major Projects," United States Government Accountability Office, GAO-17-303SP (May 2017): 49–50.

# CHAPTER 9. TREES AND ARCHAEOLOGICAL TREASURES

1. W. B. Krabill et al., "Airborne Laser Topographic Mapping Results from Initial Joint NASA/U.S. Army Corps of Engineers Experiments," NASA Technical Memorandum 73287 (June 1980): 24, https://ntrs.nasa.gov/search.jsp?R=19800019281.

2. Matti Maltamo, Erik Naesset, and Jari Vauhkonen, eds., *Forestry Applications of Airborne Laser Scanning: Concepts and Case Studies* (London: Springer, 2014), 3.

3. Jon Ranson, "Overview of Vegetation Structure Missions" (Paper presented at the Vegetation Structure Workshop, Charlottesville, VA, March 3–5, 2008), https://cce.nasa.gov/veg3dbiomass/ransontalk.pdf; Michael Lefsky, interview, September 18, 2017.

4. Gordon Maclean, telephone interview, January 28, 2018.

5. Ibid.

6. Ross Nelson, William Krabill, and Gordon Maclean, "Determining Forest Canopy Characteristics Using Airborne Lidar Data," *Remote Sensing of Environment* 15, no. 3 (June 1984).

7. William B. Krabill and Robert N. Swift, "Airborne Lidar Experiments at the Savannah River Plant," NASA Technical Memorandum 4007 (1987): 34–39.

8. Maclean, interview.

9. Krabill and Swift, "Airborne Lidar Experiments," 35.

10. Full-waveform lidar is not to be confused with coherent lidar of the sort Milton Huffaker invented to measure wind speed based on the Doppler effect of particles flying around with the wind. Full-waveform lidar is incoherent, which I hope this explanation isn't.

11. Frank E. Hoge et al., "Airborne Lidar Detection of Subsurface Oceanic Scattering Layers," *Applied Optics* 27, no. 19 (October 1, 1988).

12. Others were dabbling in airborne full-waveform lidar over forests. One notable early effort was sponsored by Canadian Forestry Service, which in summer 1983 flew the Canada Centre for Remote Sensing's Mark II bathymetric lidar over forests in Ontario. See A. H. Aldred and G. M. Bonnor, "Application of Airborne Lasers to Forest Surveys," Information Report PI-X-51, Petawawa National Forestry Institute, 1985.

13. James B. Garvin et al., "High-Resolution Measurements of Surface Topography with Airborne Laser Altimetry and Global Positioning System" (Paper presented at the NASA Fourth Airborne Geoscience Workshop, Washington, DC, January 1, 1991), 145–46, https://ntrs.nasa.gov/search.jsp?R=19910016148.

14. John Weishampel, telephone interview, September 20, 2017.

15. Hank Shugart, telephone interview, October 9, 2017.

16. M. A. Lefsky et al., "Lidar Remote Sensing of the Canopy Structure and Biophysical Properties of Douglas-Fir Western Hemlock Forests," *Remote Sensing of Environment* 70, no. 3 (December 1999); Michael Lefsky, "Research—Previous," Lefsky.org, accessed January 28, 2018, https://lefsky.org/research-previous/.

17. LVIS stands for either "Land, Vegetation, and Ice Sensor" or "Laser Vegetation Imaging Sensor," depending on whom you ask. Goddard is still flying a much-evolved version of it as part of Operation IceBridge and elsewhere. It sends out one hundred 1064-nanometer beams across a two-kilometer-wide swath. See "About LVIS," Goddard Space Flight Center, https://lvis.gsfc.nasa.gov/.

18. Sassan S. Saatchi et al., "Benchmark Map of Forest Carbon Stocks in Tropical Regions across Three Continents," *Proceedings of the National Academy of Sciences* 108, no. 24 (June 14, 2011). REDD, established in 2008, stands for Reduced Emissions from Deforestation and Forest Degradation (see www.un-redd.org/). A more recent paper explains why we should care: 35 percent of preindustrial forest cover is gone. Of what remains, as much as 82 percent is degraded from logging, urbanization, agriculture, and such. In addition to contributing to climate change (cutting/weakening forests releases aboveground and belowground carbon), forest degradation can change weather patterns, cut biodiversity, limit water resources, increase fire risk, and more.

19. James E. M. Watson et al., "The Exceptional Value of Intact Forest Ecosystems," *Nature Ecology & Evolution* 2 (February 26, 2018), doi:10.1038/s41559-018-0490-x.

20. David Gwenzi et al., "Prospects of the ICESat-2 Laser Altimetry Mission for Savanna Ecosystem Structural Studies Based on Airborne Simulation Data," *ISPRS Journal of Photogrammetry and Remote Sensing* 119 (2016). As compared to ICESat-2's ability to measure ice elevation, which should be more than twice as sharp as the original ICESat's. Despite having much higher resolution, ICESat-2's green laser light gets largely absorbed by leaves, diminishing an already weak signal.

21. Lefsky, interview.

22. Radar is lower resolution but would provide a wider swath per orbit, enabling more frequent total coverage, and it can also see through clouds.

23. Scott Goetz, "The Lost Promise of DESDynI," *Remote Sensing of Environment* 115, no. 11 (November 15, 2011).

24. Such a clean room allows for max ten thousand particles per cubic meter of air, which sounds like a lot until you consider that tens of millions of particles per cubic meter are fluttering about what you're breathing as you read this (unless you're reading this in a clean room).

25. Xcel Energy – In Your Community, "Xcel Energy's Vegetation Management: Challenges & Solutions," YouTube video, 16:49, November 7, 2013, accessed March 14, 2018, https://www.youtube.com/watch?v=khSBGoYMeDo&t=66s.

26. US-Canada Power System Outage Task Force, *Final Report on the August 14, 2003, Blackout in the United States and Canada: Causes and Recommendations* (Washington, DC: US Department of Energy, April 2004), 46, https://emp.lbl.gov/publications/final-report-august-14-2003-blackout.

27. Jim Downie, telephone interview, March 6, 2018.

28. NERC's Vegetation Management Standard, FAC-003-1, went into effect in June 2007 and has been revised three times since.

29. As of March 2018, power lines hadn't been formally established as the culprit, but lines touching trees during a fierce windstorm were suspected. The utility was told of sparking or falling equipment at fifty-four locations in Sonoma County alone the first night of the fires. David R. Baker, "Sonoma County to Sue PG&E over Fires," *San Francisco Chronicle*, February 1, 2017, https://www.sfchronicle.com/business/article/Sonoma-County-to-sue-PG-E-over-fires-12544606.php.

30. Gene Roe, "Industry Pioneers: Alastair Jenkins, CEO of GeoDigital International," *Lidar Magazine*, April 12, 2013, www.lidarmag.com/content/view/9641/198/.

31. Don Carswell, email message to author, March 7, 2018.

32. Downie, interview.

33. Chuck Anderson, telephone interview, March 6, 2018; Scott Rogers, telephone interview, March 6, 2018.

34. Gregory Berghoff, telephone interview, January 30, 2018.

35. "Über uns," TopScan GmbH, accessed January 30, 2018, www.topscan.de/front_content.php?idcat=3; Allan Carswell, email message to author, January 30, 2018; Don Carswell, email message to author, January 30, 2018.

36. Berghoff, interview.

37. David J. Harding and Gregory S. Berghoff, "Fault Scarp Detection beneath Dense Vegetation Cover: Airborne Lidar Mapping of the Seattle Fault Zone, Bainbridge Island, Washington State" (Paper presented at the American Society for Photogrammetry and Remote Sensing, June 22–26, 2000), 2–3, https://ntrs.nasa.gov/search.jsp?R=20000060844.

38. Airborne Laser Mapping had closed its doors by then. Terrapoint, a commercial spin-off of the Houston Advanced Research Center (HARC), took over the mapping. HARC had developed its own lidar, based on the ATM Bill Krabill's team had developed, with technology-transfer funding from NASA. Ibid., 8.

39. Robert Gutierrez et al., "Precise Airborne Lidar Surveying for Coastal

Research and Geohazards Applications" (Paper presented at the International Archives of Photogrammetry and Remote Sensing, 34-3/W4, Annapolis, MD, October 22–24, 2001). The US Geological Survey sponsored the work.

40. Ibid., 188.

41. R. H. Bewley, S. P. Crutchley, and C. A. Shell, "New Light on an Ancient Landscape: Lidar Survey in the Stonehenge World Heritage Site," *Antiquity* 79 (2005): 645. There's another, much earlier archaeology-focused lidar mission to note: in 1984, on the way back from a mission over waters off Chile to cross-check data from SIR-B, a space-shuttle-based radar, the NASA Wallops Airborne Oceanographic Lidar overflew a site in Costa Rica for researchers Payson Sheets and Tom Sever. In a write-up they did four years after the fact, they mention it only briefly: "As the lidar passes over an eroded footpath that still affects the topography, the pathway's indentation is recorded by the laser beam." See Payson Sheets and Tom Sever, "High-Tech Wizardry," *Archaeology* 41, no. 6 (November/December 1988): 30.

42. Jason B. Drake et al., "Estimation of Tropical Forest Structural Characteristics Using Large-Footprint Lidar," *Remote Sensing of Environment* 79, no. 2–3 (February 2002); Weishampel, interview.

43. Patricia Daukantas, "Adding a New Dimension: Lidar and Archaeology," *Optics & Photonics News*, January 2014: 36.

44. Weishampel, interview.

45. Arlen Chase, telephone interview, September 21, 2017.

46. Ibid.

47. Arlen F. Chase et al., "Airborne LiDAR, Archaeology and the Ancient Maya Landscape at Caracol, Belize," *Journal of Archaeological Science* 38, no. 2 (February 2011): 391; Chase, interview.

48. Chase, "Airborne LiDAR," 391.

49. Originally published as a supplemental figure in the online version of Chase, Chase, and Weishampel, "Lasers in the Jungle."

50. Arlen F. Chase, Diane Z. Chase, and John F. Weishampel, "Lasers in the Jungle," *Archaeology* 63, no. 4 (July/August 2010), https://archive.archaeology.org/1007/etc/caracol.html.

51. See Douglas Preston, "The Eldorado Machine," *New Yorker*, May 6, 2013. Preston's 2017 book, *The Lost City of the Monkey God*, follows the adventure and a 2015 follow-up.

52. Tom Clynes, "Laser Scans Reveal Maya 'Megalopolis' below Guatemalan Jungle," *National Geographic News*, February 1, 2018, https://news.nationalgeographic.com/2018/02/maya-laser-lidar-guatemala-pacunam/.

## CHAPTER 10. GLASS HALF FULL

1. "Got Snow? The Need to Monitor Earth's Resources," NASA, accessed February 9, 2018, https://snow.nasa.gov/sites/default/files/Got_SnowSM.pdf; "California Agricultural Production Statistics," California Department of Food & Agriculture, accessed February 9, 2018, https://www.cdfa.ca.gov/statistics/.

2. An acre foot is 325,851 gallons, which is about what five residential customers use in a year. It's a football field without the end zones that's a foot deep or a baseball infield that's five feet deep.

3. Joe Busto, telephone interview, August 1, 2017.

4. Thomas Painter, telephone interview, August 22, 2017.

5. Ibid.

6. The Carnegie Airborne Observatory, with much-upgraded hardware, is still flying over tropical forests. See its website at https://cao.carnegiescience.edu/.

7. The lidar data were, as Deems described it, an "ancillary dataset" of NASA's 2001–2004 Cold Land Process Field Experiment. See http://www.nohrsc.noaa.gov/~cline/clpx.html.

8. The model, first called WRF-Hydro and now known as the National Weather Model, was developed by an NCAR group led by David Gochis. See https://nar.ucar.edu/2017/ral/wrf-hydro-and-national-water-model.

9. Painter, interview.

## CHAPTER 11. FIREPOND

1. It's not just the US military: defense establishments in many other countries have supported lidar. For a few examples, see Vasil Molebny, Paul McManamon, Ove Steinvall, Takao Kobayashi, and Weibiao Chen, "Laser Radar: Historical Prospective—from the East to the West," *Optical Engineering* 56, no. 3 (December 2016); Vasyl Molebny, Gary Kamerman, and Ove Steinvall, "Laser Radar: From Early History to New Trends," *Proceedings of SPIE* 7835 (2010); and Paul McManamon, Gary Kamerman, and Milton Huffaker, "A History of Laser Radar in the United States," *Proceedings of SPIE* 7684 (2010).

2. John T. Correll, "The Emergence of Smart Bombs," *Air Force Magazine*, March 2010.

3. Molebny et al., "Laser Radar: Historical Prospective," 4. GPS-guided smart bombs came to the fore after the first Gulf War, but the proliferation of GPS-jamming hardware could be leading to lasers on GPS-guided bombs.

See Joe Gould, "Guided-Bomb Makers Anticipate GPS Jammers," *Defense News*, May 31, 2015, https://www.defensenews.com/air/2015/05/31/guided-bomb-makers-anticipate-gps-jammers/.

4. Gary Kamerman, telephone interview, July 1, 2017. Note that the military tends to use the term laser radar or "ladar" for lidar. Historically, "lidar" pertained to atmospheric uses and "ladar" to ranging or the discrimination of objects. With time, these distinctions have blurred. We'll go with "lidar."

5. Kamerman, interview.

6. See Kenneth P. Werrell, *The Evolution of the Cruise Missile* (Maxwell AFB, AL: Air University Press, 1985).

7. Kamerman, interview; Ronald Kaehr, email message to author, June 30, 2017; Paul McManamon, telephone interview, June 22, 2017.

8. Kaehr, email.

9. Kamerman, interview; "General Dynamics AGM-129A Advanced Cruise Missile," National Museum of the US Air Force, May 29, 2015, http://www.nationalmuseum.af.mil/Visit/Museum-Exhibits/Fact-Sheets/Display/Article/198028/general-dynamics-agm-129a-advanced-cruise-missile/.

10. For 360-degree coverage, you need multiple phased arrays. If the swim noodle analogy doesn't work for you, check out Wikipedia's page on phased arrays, which has some good graphics: https://en.wikipedia.org/wiki/Phased_array.

11. McManamon, interview.

12. Ibid.; "AN/FPS-108 'Cobra Dane,'" *Radartutorial.eu*, accessed February 14, 2018, http://www.radartutorial.eu/19.kartei/01.oth/karte003.en.html.

13. Paul F. McManamon, "Optical Phased Array Technology," *Proceedings of the IEEE* 84, no. 2 (February 1996).

14. McManamon, interview; John Carrano, "Increasing the Effectiveness of Steered Agile Beams," DARPA, accessed February 14, 2018, http://archive.darpa.mil/DARPATech2002/presentations/mto_pdf/speeches/CARRANO.pdf.

15. Frederik Nebeker, "Oral History: Leo Sullivan," IEEE History Center, June 14, 1991, http://ethw.org/Oral-History:Leo_Sullivan; Brian Edwards, telephone interview, November 3, 2017.

16. Alan A. Gromenstein, ed., *MIT Lincoln Laboratory: Technology in Support of National Security* (Lexington, MA: MIT, 2011), 409. Firepond sounds like an evocative name for a high-powered laser system; turns out that the MIT/LL Millstone Hill site lacked fire-hydrant-class water supply, so they had dug a big hole and allowed it to fill with water: so a pond in case of fire. Millstone Hill, for the record, was named after old millstones they found when setting up Firepond cohabitants such as the giant Haystack antenna, which was named after the place

a farmer once stored his hay. Firepond's official name was the Firepond Optical Research Facility.

17. The LRPA (laser radar power amplifier) was to increase the milk-tube output of one thousand watts to a peak power of two hundred thousand watts. Edwards, interview; Gromenstein, *MIT Lincoln Laboratory*, 410–11; Alfred B. Gschwendtner and William E. Keicher, "Development of Coherent Laser Radar at Lincoln Laboratory," *Lincoln Laboratory Journal* 12, no. 2 (2000): 384–85.

18. Edwards, interview. Edwards says the problem with gas circulating within the LRPA units. The gas was hammered through by thousand-horsepower motors typically used for drying grain, and the gas scraped up chips of paint from inside LRPA's painted-steel piping. The paint chips punched holes in the foil of an electron gun, which ruined the vacuum the LRPA needed to do its job. The program was canceled by the time they solved it and imaged a tumbling Agena D rocket booster from 1,350 kilometers away in 1981. See Gschwendtner, "Development of Coherent Laser Radar," 385.

19. Ronald Reagan, "Address to the Nation on Defense and National Security," Ronald Reagan Presidential Library and Museum, March 23, 1983, https://www.reaganlibrary.gov/sites/default/files/archives/speeches/1983/32383d.htm.

20. Edwards, interview.

21. This was based on a radar technique called coherent wideband radar, which circumvents the conventional angular resolution limits imposed by diffraction and atmospheric turbulence. Gromenstein, *MIT Lincoln Laboratory*, 411.

22. A stable laser is one whose frequency and power varies as little as possible. With a Doppler detector, that's particularly important because to detect something that's moving, you have to trust the original signal you're comparing the return signal to.

23. The gas was a rare carbon dioxide isotope that cost $80,000 to refill, Edwards says.

24. Edwards, interview. Gromenstein, *MIT Lincoln Laboratory*, 413–14. Spectra Technology, a contractor, built these beastly amps under the tutelage of Firepond veteran Lincoln Lab scientist Charles Freed.

25. Edwards, interview.

26. Ibid.; Gromenstein, *MIT Lincoln Laboratory*, 413–14.

27. On the scientific side, the most potent skepticism arrived with the "Report to the American Physical Society of the Study Group on Science and Technology of Directed Energy Weapons," *Review of Modern Physics* 59, no. 3, part 2 (July 1987).

28. Ken Adelman, "The Phantom Menace," *Politico*, May 11, 2014, https://www.politico.com/magazine/story/2014/05/the-phantom-menace-106551.

29. Ibid.

# CHAPTER 12. MAP QUEST

1. Richard Marino, telephone interview, November 11, 2017.

2. Paul McManamon, telephone interview, June 22, 2017. The Firebird test used photomultipliers rather than Geiger-mode detectors, which were still considered too risky for an expensive experiment, Marino says. For more on the technical side of all this, see Marius A. Albota et al., "Three-Dimensional Imaging Laser Radars with Geiger-Mode Avalanche Photodiode Arrays," *Lincoln Laboratory Journal* 13, no. 2 (2002).

3. Brian F. Aull et al., "Geiger-Mode Avalanche Photodiodes for Three-Dimensional Imaging," *Lincoln Laboratory Journal* 13, no. 2 (2002): 342–43.

4. Marino, interview.

5. Albota, "Three-Dimensional Imaging Laser Radars," 355, 358. Q-switching refers to the way pulses are created. Zayhowski ran an 808-nanometer diode laser's output through optical fiber into his new laser. The second crystal (chromium-doped YAG) served as a saturable absorber that the onset of lasing in the Nd:YAG cavity saturated, essentially opening the gate for a laser pulse and then closing it again.

6. This was one of what would be a dizzying number of applications for the technology. Zayhowski reached out to military, scientific organizations and industry to drum up interest. Within a few years, it was licensed to thirteen organizations for use in 3D imaging, semiconductor micromachining, biological warfare agent detection, remote chemical detection, and microsurgery, among other areas. John Zayhowski, email message to author, October 4, 2017; John J. Zayhowski, "Passively Q-Switched Microchip Lasers—A Historical Perspective," MIT Lincoln Laboratory, June 1, 2010.

7. Marino, interview. The first Geiger-mode systems used green light, frequency-doubled Nd:YAG at 532 nanometers. ALIRT (Airborne Ladar Imaging Research Testbed) introduced indium gallium arsenide detectors that could see in the infrared. HALOE stood for High-Altitude Lidar Operations Experiment.

8. National Research Council, *Laser Radar: Progress and Opportunities in Active Electro-Optical Sensing* (Washington, DC: National Academies Press, 2014), 40; David Walsh, "Laser-Based Mapping Tech a Boost for Troops in Afghanistan," *GCN*, July 27, 2011, https://gcn.com/articles/2011/07/18/tech-watch-geoint-lidar.aspx.

9. ALIRT fed into a more powerful, higher-flying version called MACHETE. Jigsaw fed into JAUDIT (Jungle Advanced Under Dense Vegetation Imaging Technology), which was then renamed TACOP (Tactical Operational

Lidar). This was as of 2013; they've continued to push ahead, Marino says. "The technology is actually more capable than we've described," he says. "There are new advances we haven't talked about at all, and they're not published anywhere."

10. John Degnan, interview, October 26, 2017.

11. The Microlaser Altimeter pulsed 3,800 times per second at 7.6 milliwatts per pulse. John Degnan, email message to author, March 17, 2018.

12. Degnan, interview; John J. Degnan, "Scanning, Multibeam, Single Photon Lidars for Rapid, Large Scale, High Resolution, Topographic and Bathymetric Mapping." *Remote Sensing* 8, no. 11 (2016): 3–4.

13. "BuckEye," *GlobalSecurity.org*, accessed February 19, 2017, https://www.globalsecurity.org/intell/systems/buckeye.htm; Mike Hardaway, "BuckEye in 2011 and Beyond: Delivering High-Resolution Terrain Data to the Warfighter," *LiDAR Magazine*, October 17, 2011, www.lidarmag.com/content/view/8646/198/; Degnan, "Scanning, Multibeam, Single Photon," 7–8.

14. Johannes Riegl Jr., email message to author, January 4, 2018.

15. UAS stands for Unmanned Aerial System, which seems to be the preferred term these days among specialists, taking the crown from Unmanned Aerial Vehicle (UAV). "Drone" works for short. The FAA rule, 14 CFR part 107, can be found at https://www.faa.gov/uas/getting_started/part_107/.

16. Chuck Anderson, telephone interview, March 6, 2018.

17. Leica Geosystems' SPL 100 lidar has a ten-by-ten focal plane array as a sensor; the Harris Geiger-mode lidar has a 128-by-32-pixel array made by Princeton Lightwave.

18. Hope Morgan, interview, February 6, 2018.

19. "John Wesley Powell," USGS, accessed February 23, 2018, https://www.usgs.gov/staff-profiles/john-wesley-powell.

20. Jason Stoker, "The 3D Elevation Program: Overview" (Presentation, 2015), http://dels.nas.edu/resources/static-assets/besr/miscellaneous/MSC/2015/3DEP_Stoker.pdf; R. T. Evans and H. M. Frye, "History of the Topographic Branch (Division)," *US Geological Survey Circular* 1341 (2009): 3–4, 25–28.

21. Larry J. Sugarbaker et al., "The 3D Elevation Program Initiative—A Call for Action," *US Geological Service Circular* 1399 (2014): 7.

22. Mary Williams Walsh, "A Broke, and Broken, Flood Insurance Program," *New York Times*, November 5, 2017.

23. "S.698 – National Landslide Preparedness Act," Congress.gov, accessed February 23, 2018, https://www.congress.gov/bill/115th-congress/senate-bill/698.

24. Sugarbaker, "3D Elevation Program," 6.

## CHAPTER 13. ATOMS TO BYTES

1. Ben Kacyra, interview, September 26, 2017.

2. Marc Cheves, "Monumental Challenge," *American Surveyor*, November/December 2014, http://www.amerisurv.com/PDF/TheAmericanSurveyor_Cheves-MonumentalChallenge_Nov-Dec2014.pdf.

3. Kacyra, interview. Engineers use a CAD, or computer-aided design, drawing to design things.

4. Ibid.

5. John Zayhowski, email messages to author, October 4, 2017, and March 1, 2018; Kacyra, interview.

6. Gene Roe, "Industry Pioneers: Jerry Dimsdale," *LIDAR Magazine*, September 5, 2014, http://www.lidarmag.com/content/view/10906/.

7. A total station is the instrument on a tripod you've seen surveyors peering through. The modern ones use lasers, but they take one shot at a time, such that a skilled surveyor might get a few hundred readings done a day. Laser scanners like the one Cyra Technologies pioneered now capture on the order of a million shots per second.

8. Kacyra, interview; Roe, "Jerry Dimsdale."

9. Independently, Riegl delivered its first 3D scanner to an OEM customer in 1998; that became the LMS-Z210, which sold commercially starting a year or so later. Riegl, email.

10. Ben Kacyra, "Overview of 3D Scanning (HDS) and its Applications in the AEC Market" (Presentation, 2002).

11. Kacyra, interview.

12. John Ristevski, telephone interview, October 10, 2017.

13. Mark Harris, "God Is a Bot and Anthony Levandowski Is His Messenger," *Wired*, September 27, 2017, https://www.wired.com/story/god-is-a-bot-and-anthony-levandowski-is-his-messenger/. 510 Systems also developed a similar system for Navteq, called the Navteq True.

14. Here provides the data for Bing Maps; Google provides its own for Google Maps. These weren't the first or only mobile mapping efforts—just the biggest ones. One early example was in Afghanistan in 2003, where the Canadian company Mosaic Mapping Solutions mounted an airborne lidar on a pole in a pickup truck bed to map 350 miles of highway between Kabul and Kandahar. The three employees were accompanied by fifteen to twenty security personnel. See Simon Newby and Paul Mrstik, "LiDAR on the Level in Afghanistan," *GPS World*, July 2005 (reprint). Mosaic was led by Alastair Jenkins, who would later apply lidar

to vegetation management for Xcel Energy and others with his follow-up company, GeoDigital.

15. Check it out at https://artsandculture.google.com/project/cyark.

16. As much as I'd like not to yell "FARO," I let "earthmine" get away without a capital and feel it's only fair.

17. Simon Raab, telephone interview, November 28, 2017.

18. Ibid.; "FARO Technologies, Inc.," *International Directory of Company Histories*, Encyclopedia.com, accessed February 28, 2018, http://www.encyclopedia.com/books/politics-and-business-magazines/faro-technologies-inc.

19. Raab, interview.

20. Michael Raphael, interview, October 26, 2017.

21. Raab, interview. Raab stayed on as CEO until 2006, remaining as chairman until taking the reins again in 2015. It's worth noting that Raab is as well-known as a contemporary artist as he is as a scientist/technologist/entrepreneur. He invented and patented a technique he calls Parleau, derived from the French "par l'eau," for "through water." He works with special translucent paints on layers of polymer adhered to steel or aluminum using lessons learned from surface physics and artificial joint implants. The color doesn't crack when he crumples the metal. See simonraabgallery.com.

22. The FARO Focus line works a bit differently than the Cyrax/Leica Geosystems machines. Rather than time-of-flight measurement, they send out a continuous infrared beam and look for the phase shift of the light waves coming back. Also, FARO wasn't the only company doing metrology with articulated arms. Romer, a Carlsbad, California, company, developed portable measuring arms, which were paired with a laser scanner made by Perceptron. It's now part of Leica Geosystems.

23. Raphael hasn't been the only one doing these things, he is quick to note. In Hollywood, University of Southern California computer scientist Paul Debevec has been a scanning pioneer, digitizing Brad Pitt's face for the first hour of *The Curious Case of Benjamin Button*, among other efforts. In the arts, a Stanford team led by Marc Levoy spent the 1998–1999 academic year scanning ten Michelangelo statues, including the *David*, with a custom lidar rig in Florence, Italy. See http://graphics.stanford.edu/projects/mich/.

## CHAPTER 14. HIT THE ROAD (SLOWLY)

1. Archive.org has a cool video of the Stanford Cart here: https://archive.org/details/sailfilm_cart.

2. Les Earnest, "Stanford Cart," December 2012, https://web.stanford

.edu/~learnest/cart.htm; Robert Reinhold, "'Baby Robot Learns to Navigate in a Cluttered Room," *New York Times*, April 10, 1968; Hans Moravec, "Robots That Rove," *Omni*, September 11, 1984, http://www.frc.ri.cmu.edu/~hpm/project.archive/general.articles/1985/1985.Omni.html.

3. Hans Moravec, "Obstacle Avoidance and Navigation in the Real World by a Seeing Robot Rover" (PhD diss., Stanford University, 1980), http://www.frc.ri.cmu.edu/users/hpm/project.archive/robot.papers/1975.cart/1980.html.thesis/p02.html. This as entertaining as it gets for a PhD dissertation. The opening line: "This is a report about a modest attempt at endowing a mild-mannered machine with a few of the attributes of higher animals."

4. Moravec, "Robots That Rove"; Earnest, "Stanford Cart."

5. Moravec, "Robots That Rove."

6. Red Whittaker, telephone interview, June 23, 2017.

7. Ibid.

8. Red Whittaker, interview, October 24, 2017; Christine O'Toole, "What Drives Red Whittaker?," *Pittsburgh Quarterly*, Winter 2018, https://pittsburghquarterly.com/pq-health-science/pq-technology/item/1566-what-drives-red-whittaker.html.

9. Whittaker, interview, October 24, 2017.

10. Daniel Lovering, "Radioactive Robot: The Machines that Cleaned Up Three Mile Island," *Scientific American*, March 27, 2009, https://www.scientificamerican.com/article/three-mile-island-robots/.

11. Whittaker, interview, October 24, 2017.

12. Lee Champeny-Bares, Syd Coppersmith, and Kevin Dowling, "The Terregator Mobile Robot," Field Robotics Center, Carnegie Mellon University, May 1991, 10, https://www.ri.cmu.edu/pub_files/pub1/champeny_bares_1991_1/champeny_bares_1991_1.pdf.

13. Hans P. Moravec and Alberto Elfes, "High Resolution Maps from Wide Angle Sonar," The Robotics Institute, Carnegie Mellon University (July 1984), https://www.frc.ri.cmu.edu/~hpm/project.archive/robot.papers/1985/al2.html.

14. Ibid. The French HILARE autonomous robot, built starting in 1977, also used ultrasound sensors, plus a camera and a laser range finder. See "1977-'HILARE' Autonomous Mobile Robot-French," Cyberneticzoo.com, accessed March 18, 2018, http://cyberneticzoo.com/cyberneticanimals/1977-hilare-autonomous-mobile-robot-french/.

15. See "Strategic Computing," DARPA (October 28, 1983): i, v, www.dtic.mil/dtic/tr/fulltext/u2/a141982.pdf.

16. Takeo Kanade and Charles Thorpe, "CMU Strategic Computing Vision

Project Report: 1984 to 1985," The Robotics Institute, Carnegie Mellon University (November 1985): 31, http://repository.cmu.edu/cgi/viewcontent.cgi?article=1554&context=robotics.

17. Ron Swonger, telephone interview, September 12, 2017.

18. Dwight Carlson, telephone interview, September 25, 2017.

19. Paul F. McManamon, Gary Kamerman, and Milton Huffaker, "A History of Laser Radar in the United States," *Proceedings of SPIE* 7684 (April 2010): 4.

20. C. D. Miller, "Final Report: Image Sensor Data Base for the DARPA ALV Program," ERIM, October 1986, http://www.dtic.mil/dtic/tr/fulltext/u2/a178117.pdf. Martin Marietta was a big Strategic Computing Initiative contractor working on the Autonomous Land Vehicle program of which Carnegie Mellon was also a part. Martin built an eight-foot-high, ten-wheeled, 15,200-pound vehicle that looked like something the Galactic Empire might carry stormtroopers in. It was filled with electronics, and the ERIM lidar peered out its sole window as it crawled about Martin's complex in Littleton, Colorado. See Jim Schefter, "Look Ma! No Driver," *Popular Science*, October 1985.

21. Miller, "Final Report," 11. The ERIM worked on the same principle as the FARO Focus tripod scanner CyArk uses today. Rather than time-of-flight measurement, it sent out a continuous infrared beam and looks for the phase shift of the light waves coming back. It was called "AM lidar" (for amplitude modulation) at the time.

22. Whittaker, interview, June 23, 2017.

23. Kevin Dowling et al., "NAVLAB: An Autonomous Navigation Testbed," The Robotics Institute, Carnegie Mellon University (November 1987), https://www.ri.cmu.edu/pub_files/pub3/dowling_kevin_1987_1/dowling_kevin_1987_1.pdf.

24. Whittaker, interview, June 23, 2017.

25. Sanjiv Singh, interview, October 23, 2017.

26. Charles Thorpe et al., "Toward Autonomous Driving: The CMU Navlab," *IEEE Expert* 6, no. 4 (August 1991): 37. The Perceptron had its share of issues too, having a hard time staying calibrated if it got warmer than about eighty degrees, suffering range drift depending on surface materials, and beam divergence because of diffraction from the beam telescope. In So Kweon, Regis Hoffman, and Eric Krotkov, "Experimental Characterization of the Perceptron Laser Rangefinder," The Robotics Institute, Carnegie Mellon University (June 1991).

27. Whittaker, interview, June 23, 2017.

28. Sanjiv Singh and Jay West, "Cyclone: A Laser Scanner for Mobile Robot Navigation," The Robotics Institute, Carnegie Mellon University (September 1991): 3, 6, 18; Singh, interview, October 23, 2017.

29. Singh and West, "Cyclone," 16.

30. Todd Jochem, "They Drove Cross-Country in an Autonomous Minivan without GPS. In 1995," *Jalopnik*, April 9, 2015, https://jalopnik.com/they-drove-cross-country-in-an-autonomous-minivan-witho-1696330141. This was not the first autonomous highway vehicle. Others include the Tsukuba Mechanical Engineering Lab's driverless car in 1977 (using cameras, it could track white street markers with machine vision) and Ernst Dickmann's VaMoRs Mercedes van, developed from 1986–2003 as part of the European Eureka PROMETHEUS project. See Marc Weber, "Where To? A History of Autonomous Vehicles," Computer History Museum, May 8, 2014, http://www.computerhistory.org/atchm/where-to-a-history-of-autonomous-vehicles/.

31. Erik Krotkov et al., "Ambler: A Six-Legged Planetary Rover," *IEEE Advanced Robotics* (1991), doi:10.1109/ICAR.1991.240684.

32. For more on these and many other projects, see "Robots at the FRC," Field Robotics Center, Carnegie Mellon, http://www.frc.ri.cmu.edu/robots/.

33. Anne Watzman, "Groundhog Debut," *Carnegie Mellon Today*, October 31, 2002.

## CHAPTER 15. DEBACLE IN THE DESERT

1. National Defense Authorization Act for Fiscal Year 2001, S. 2549, Sec. 217, 106th (2000), https://www.congress.gov/bill/106th-congress/senate-bill/2549/text.

2. Alex Davies, "An Oral History of the DARPA Grand Challenge, the Grueling Robot Race That Launched the Self-Driving Car," *Wired*, August 3, 2017, https://www.wired.com/story/darpa-grand-challenge-2004-oral-history/.

3. Chris Urmson et al., "High Speed Navigation of Unrehearsed Terrain: Red Team Technology for Grand Challenge 2004," The Robotics Institute, Carnegie Mellon University (June 1, 2004): 12–15.

4. Michael Darter, "DARPA's Debacle in the Desert," *Popular Science*, June 4, 2004, https://www.popsci.com/scitech/article/2004-06/darpa-grand-challenge-2004darpas-debacle-desert. Hat tip to *Popular Science* for the title of this chapter.

5. Ibid.; Davies, "Oral History."

6. David Hall, interview, September 27, 2017.

7. Ibid.

8. David S. Hall, "Team D.A.D. Technical Paper," DARPA Grand Challenge 2004 (February 19, 2004), http://archive.darpa.mil/grandchallenge05/TechPapers/TeamDAD.pdf.

9. Jim McBride, interview, August 10, 2017.

10. A film crew for the PBS series *NOVA* showed up with them, chronicling for what would run as "The Great Robot Race." Note that at the beginning of the clip, Bruce Hall asks his brother, "How many lasers again?" to which David answers, "Sixty-four." The printed circuit board with the Texas Instruments DSP on it that Hall is shown working on at a microscope was destined for his new lidar. See "DAD's Big Day," video, *NOVA*, 2:30, accessed March 28, 2018, http://www.pbs.org/wgbh/nova/darpa/outtakes.html.

11. See "Reaction Time Statistics," Human Benchmark, https://www.humanbenchmark.com/tests/reactiontime/statistics.

12. These were Osram 905-nanometer diode lasers; the detectors came from National Semiconductor.

13. Chris Urmson et al., "Tartan Racing: A Multi-Modal Approach to the DARPA Urban Challenge," DARPA Grand Challenge 2007 (April 13, 2007): 20–21, http://archive.darpa.mil/grandchallenge/TechPapers/Tartan_Racing.pdf.

14. Stanford Racing Team, "Stanford's Robotic Vehicle 'Junior': Interim Report," DARPA Grand Challenge 2007 (April 12, 2007), http://archive.darpa.mil/grandchallenge/TechPapers/Stanford.pdf.

## CHAPTER 16. THE ROAD AHEAD

1. It was a big deal at the time, though, pitting "the most successful company from the dot-com era [that is, Google, whose parent Alphabet also owns Waymo] against this generation's biggest start-up in a fight over autonomous vehicles—a potentially trillion-dollar industry that is expected to transform transportation," as the *New York Times* put it. It was about lidar too: in particular, about Anthony Levandowski, who had entered the motorcycle in the 2004 DARPA Grand Challenge and later was a Google autonomous vehicle program leader. He allegedly stole lidar technology that Google/Waymo had spent years and tens of millions of dollars developing before leaving to found autonomous trucking company Otto, which Uber promptly bought. The case was settled in February 2018, with Uber forking over 0.34 percent of its stock to Waymo, valued at $245 million. Daisuke Wakabayashi, "Uber and Waymo Settle Trade Secrets Suit over Driverless Cars," *New York Times*, February 9, 2018.

2. Daisuke Wakabayashi, "Uber's Self-Driving Cars Were Struggling Before Arizona Crash," *New York Times,* March 23, 2018.

3. "Critical Reasons for Crashes Investigated in the National Motor Vehicle

Crash Causation Survey," National Highway Traffic Safety Administration, DOT HS 812 115 (February 2015).

4. NHTSA Public Affairs, "USDOT Releases 2016 Fatal Traffic Crash Data," October 6, 2017, https://www.nhtsa.gov/press-releases/usdot-releases-2016-fatal-traffic-crash-data.

5. National Safety Council, "NSC Motor Vehicle Fatality Estimates," accessed March 25, 2018, http://www.nsc.org/NewsDocuments/2017/12-month-estimates.pdf.

6. World Health Organization, "Number of Road Traffic Deaths," accessed March 29, 2018, www.who.int/gho/road_safety/mortality/traffic_deaths_number/en/.

7. Paul Barter, "Cars Are Parked 95% of the Time. Let's Check!" *Reinventing Parking* (blog), February 22, 2013, https://www.reinventingparking.org/2013/02/cars-are-parked-95-of-time-lets-check.html.

8. Charlie Johnson and Jonathan Walker, "Peak Car Ownership: The Market Opportunity of Electric Automated Mobility Services," Rocky Mountain Institute, September 2016, 5, https://rmi.org/insights/reports/peak-car-ownership-report/.

9. Shaller Consulting, "Unsustainable? The Growth of App-Based Ride Services and Traffic, Travel and the Future of New York City," accessed March 29, 2018, http://schallerconsult.com/rideservices/unsustainable.htm.

10. Michele Bertoncello and Dominik Wee, "Ten Ways Autonomous Drive Could Redefine the Automotive World," McKinsey & Company, accessed March 25, 2018, https://www.mckinsey.com/industries/automotive-and-assembly/our-insights/ten-ways-autonomous-driving-could-redefine-the-automotive-world.

11. Aric Jenkins, "GM Wants to Bring an Uber-Like Self-Driving Car Service to Big Cities in 2019. Will It Work?," *Fortune*, November 30, 2017, http://fortune.com/2017/11/30/gm-autonomous-ride-share-2019/.

12. Chunka Mui, "How Kodak Failed," *Forbes*, January 18, 2012, https://www.forbes.com/sites/chunkamui/2012/01/18/how-kodak-failed/

13. Waymo LLC v. Uber Technologies, 17-cv-00939-WHA (N.D. Calif. 2017), https://www.courttrax.com/wp-content/uploads/2017/03/USDC-CA-N-3-17cv00939-Complaint.pdf.

# CHAPTER 17. GLUE, DRONES, AND RADIOACTIVE PIPES

1. Police Radar Information Center, "Lidar Operations," http://copradar.com/chapts/chapt5/ch5d2.html; and "Antenna Beam," http://copradar.com/chapts/chapt3/ch3d2.html, both accessed April 4, 2018.

2. "Best of What's New," *Popular Science*, December 1990, 74; Mike Phippen, "LTI Historical Moment—The LTI 20/20 Marksman," *Traffic Safety Blog*, Laser Technology, June 6, 2012, http://www.lasertech.com/blogs/Traffic-Safety/post/12/06/06/LTI-Historical-Moment-The-LTI-2020-Marksman.aspx. Police also increasingly use lidar scanners for capturing crime scenes and accident scenes, enabling greater accuracy and, in the case of accidents, faster cleanup and reopening of roads.

3. Vasil Molebny, Paul McManamon, Ove Steinvall, Takao Kobayashi, and Weibiao Chen, "Laser Radar: Historical Prospective—from the East to the West," *Optical Engineering* 56, no. 3 (December 2016): 7.

4. Charles Arthur, "Laser Spying: Is It Really Practical?" *Guardian*, August 22, 2013, https://www.theguardian.com/world/2013/aug/22/gchq-warned-laser-spying-guardian-offices.

5. "Optical Vibration Measurement per Laser Vibrometery," Polytec, accessed April 4, 2018, https://www.polytec.com/us/vibrometry/areas-of-application/. For other examples from the medical realm, see Habib Tabatabai et al., "Novel Applications of Laser Doppler Vibration Measurements to Medical Imaging," *Sensing and Imaging* 14, no. 1–2 (August 13, 2013): 13–28.

6. "Laser Vibrometry Shows How Singing Crickets Hit High Notes," *Optics.org*, May 28, 2013, http://optics.org/news/4/5/39; Jernej Polajnar et al., "Vibrational Communication of the Brown Marmorated Stink Bug (Halyomorpha halys)," *Physiological Entomology* 41 (2016).

7. USDA, "USDA-ARS Entomologists 'Turn Up the Bass' on Vineyard Pests," YouTube video, 2:15, December 7, 2017, https://youtu.be/IDivjBZ71B8.

8. Sanjiv Singh, "Synthesis of Tactical Plans for Robotic Excavation" (PhD diss., Carnegie Mellon University, 1995), 9, https://www.ri.cmu.edu/pub_files/pub1/singh_sanjiv_1995_3/singh_sanjiv_1995_3.pdf.

9. Ibid., 160.

10. Anthony Stentz et al., "A Robotic Excavator for Autonomous Truck Loading," The Robotics Institute, Carnegie Mellon University (1999), http://www.frc.ri.cmu.edu/~ssingh/pubs/truckLoading.pdf.

11. Sanjiv Singh, interview, October 23, 2017.

12. Sanjiv Singh, email message to author, April 18, 2018.

13. Lyle Chamberlain, Sebastian Scherer, and Sanjiv Singh, "Self-Aware Helicopters: Full-Scale Automated Landing and Obstacle Avoidance in Unmapped Environments" (Paper presented at the American Helicopter Society 67th Annual Forum, Virginia Beach, VA, May 2011), https://www.ri.cmu.edu/pub_files/2011/3/ahs2011_Final.pdf.

14. Garrett Hemann, Sanjiv Singh, and Michael Kaess, "Long-Range GPS-Denied Aerial Inertial Navigation with LIDAR Localization," *IEEE Intelligent Robots and Systems*, 2016, https://www.ri.cmu.edu/pub_files/2016/10/Hemann16iros.pdf.

15. Singh, email.

16. "Boeing HorizonX Invests in Unmanned Systems Technology Leader Near Earth Autonomy," press release, October 18, 2017, http://boeing.mediaroom.com/2017-10-19-Boeing-HorizonX-Invests-in-Unmanned-Systems-Technology-Leader-Near-Earth-Autonomy. Others are pursuing autonomous air taxis too. Among them, Google cofounder Larry Page's Kitty Hawk, Uber Elevate, and China's EHang.

17. Marty Reibold, telephone interview, November 10, 2018; "Take a Look Inside the Plant," Portsmouth Gaseous Diffusion Plant Virtual Museum, accessed November 15, 2017, http://portsvirtualmuseum.org/virtual-tour.htm.

18. Reibold, interview.

19. Prachi Patel, "Looking for Methane Leaks," *Chemical & Engineering News* 95, no. 35 (September 4, 2017), https://cen.acs.org/articles/95/i35/Looking-methane-leaks.html.

20. Sean Coburn et al., "Regional Trace-Gas Source Attribution Using a Field-Deployed Dual Frequency Comb Spectrometer," *Optica* 5, no. 4 (2018), doi:10.1364/OPTICA.5.000320; "Frequency Combs for Methane Detection," ARPA-E, accessed April 5, 2018, https://arpa-e.energy.gov/?q=slick-sheet-project/frequency-combs-methane-detection.

21. David Harding, interview, October 27, 2017.

22. Marie-Claire Lynn, email message to author, February 21, 2017; "GRiD: Transforming Public Spaces with Collaborative Play," Moment Factory, accessed April 5, 2018, https://momentfactory.com/lab/grid.

# INDEX